Space Security : Indian Perspective

Space Security : Indian Perspective

Gp Capt (Retd) G D Sharma, VSM

Published in association with
Centre For Joint Warfare Studies (CENJOWS)
Kashmir House, Rajaji Marg, New Delhi

Vij Books India Pvt Ltd
Darya Ganj, New Delhi (India)

Published   by

Vij Books India Pvt Ltd
(Publishers, Distributors & Importers)
2/19, Ansari Road, Darya Ganj
New Delhi - 110002
Phones: 91-11-43596460, 91-11- 65449971
Fax: 91-11-47340674
www.vijbooks.com
e-mail : vijbooks@rediffmail.com

Copyright © 2011, Centre for Joint Warfare Studies (CENJOWS),
New Delhi

ISBN: 978-93-80177-76-2

Dedicated to my late mother

Contents

FOREWORD

Space is no longer in the realm of science fiction. There is growing recognition of its impact in conducting the affairs of the nations more so in the arena of the military affairs. Revolution in Military Affairs (RMA) and Net Centric Operations are reliant on the space to some degree. Satellite communication, space derived imagery, navigation and targeting provided by the Global Positioning System are critical to the Armed Forces to deliver a specific effect at the time and place of their choosing through synchronised effect operations across the air, land, and sea. However, while we develop our space capability we must also develop redundancy so that we do not unwittingly build an "Achilles Heel" in our military hardware.

Reliance on space is not exclusive to defence alone in fact the dependence of civil sector is more critical as it would impact governance by the state and common citizen in his day to day functioning. In defence, we need not only to develop affordable options but also take steps to safeguard our space assets to ensure their continued use. Chinese Anti Satellite test in 2007 and Saddam Hussain's attempt to jam GPS in Iraq war has amply demonstrated such an imperative. While developing our response, we need to carefully balance our commitment to the peaceful use of the space under the Outer Space Treaty of 1967 of UN against our right of self defence under Article 51 of the UN Charter.

India has been developing this capability albeit for civil uses since early sixties but, its military applications are still rudimentary. We now have to factor in Space in National Security Calculus innovatively. Removal of key Indian industry from the US entity list has unwittingly provided India an opportunity to achieve greater heights with absorption of high end technology in this field. It is the inflection point and we need to enunciate a Space Doctrine and develop capability and procedures to ensure its Security.

Gp Capt. (Retd) GD Sharma has examined the impact of space in the military affairs in his study titled as "Space Security : Indian Perspective" and also suggested options to India to stay in sync in this developing field which reader may find interesting.

(KB Kapoor)
Maj Gen (Retd)
Director CENJOWS

PREFACE

The Earth first Satellite, Sputnik, was launched by erstwhile Soviet Union on 04 October 1957. In half a century thereafter, humanity has made remarkable success in the space arena and has conquered greater heights. Nearly all aspects of our daily life are impacted by the Satellites. The degree to which we rely today on space systems for many daily routine activities is quite incredible. Weather forecasting, radio and TV broadcasting, spread of education, navigation, mobile telecommunications in the high seas and remote areas, search and rescue, environment data collection etc are but a few applications which call upon satellite-based infrastructure located in Earth orbits, at altitudes ranging from a few hundreds of kilometres to the geosynchronous (GEO) orbit at 36,000 km.

As regard to use of satellites in military affairs is concerned, the recent conflicts have shown that the nations using the space based assets in war fighting gained tremendous advantage over other nations. The United States of America and its international coalition partners used satellites to their advantage in the military campaigns in Yugoslavia in 1999, Iraq in 2003 and in Afghanistan from 2001 onwards. The satellites are being extensively used by USA and its allies to seek and destroy terror outfits of AL Qaeda and Taliban in AF- Pak campaign.

Historically, military activities in outer space were an offshoot of the arms race and a corollary to the development of missile defence projects like anti-ballistic missiles (ABMs) and anti-satellites (ASATs), by the superpowers. Hence, it seems natural that these powers would have futuristic plans to exploit outer space for their national defence imperatives or security cover for their strategic allies. Not surprisingly, they have already undertaken programmes, and have more in the pipeline, for offensive and defensive systems. In either case, the risk of weaponisation of outer space is real.[1]

It is now quite clear that space assets can be used for real-time war-theatre dominance which can influence the outcome of battles. The importance of space satellites for quality surveillance, accurate pinpointing enemy positions, and transmission of data back to the earth station for analysis and conversion to command instructions by commanders in a matter of seconds is of tremendous and decisive advantage in operations. The synergy, thus, created in collaboration with earth-based systems can be amazing. Besides, speedy decision-making by the Commanders, the precise satellite aided targeting with smart munitions has also helps in minimising the collateral damage. No wonder therefore, the outer space provides the traditional high ground for observation with spectacular advantages in augmenting the military might.

India is one of the few nations that have developed the capability to fabricate its own satellites and launch them to the space at varied heights. It has launched a substantial number of satellites for communications, weather watch and earth observation. Infact, most activities of our daily life have been impacted by satellite based technologies. Recently, India also became the fifth nation to launch an unmanned moon missions. However, not so far back in time, an incident which made everybody sit up and take notice was a Chinese shooting down of a its own junk weather satellite with missile fired from one of its ground site. India has every reason to get worried over the possibility of a "Chinese Threat".Being aware of the challenges, Air Marshal FH Major, former Chief of Air Staff has clearly stated we need to make extensive use of space assets for a variety of

---

[1] A paper titled ,"Towards a future European space surveillance system "by Bhupendra Jasani in collective security in space published by space policy institute, George Washington University.

passive and active combat roles and that the IAF is fully well aware of the threat faced from "space and cyberspace". Army and naval chiefs have made similar assertions about the dimension of the threat in the backdrop of Space and nuclear capabilities of our adversaries. Even Our Defence Minister Antony deliberating after the Chinese Anti Satellite test wondered, "how long India can remain committed to the policy of non-weaponisation of space even as counter space systems are emerging in our neighbourhood." [2]

Today, militarization of the outer space seems a fait *accompli.* There are numerous satellites that have dual use for the military or civil. Global Positioning System (GPS) for example provides precise in time and positional coordinates. The service is now widely used for navigational purposes and traffic control in the air, at sea and on the ground by the civil and the military consumers. The military also uses this innocuous facility for precise targeting of enemy positions with GPS-aided weapons and aided guided munitions. GPS is aptly suited for monitoring ground troop movements as well as search and rescue missions because of its high accuracy on grid-references. Thus, a separation of exclusive military and purely civilian satellites is difficult by definition and may even be misleading. To compound the situation further, the number of mixed assets in outer space is likely to grow exponentially in the future. The stage where outer space was a protected sanctuary for scientific exploration and peaceful activities seems long over as space has been "militarised" already by both military and commercial satellites. It would therefore, be naive and futile to expect only peaceful use of space assets.

In India, our space programme, for good reasons, has been focused almost exclusively for civil uses. In the process, we have built up commendable capabilities. ISRO have charted for itself a path of scientific accomplishments mixed with societal good and has aspiration to put a

[2] Radhakrishna Rao in an article "*Chinese Threat to Indian Space Assets*" available at www.domain-b.com/aero/20090129_indian_space.html

man on the moon. But, we have failed to pay attention to global trends on the increased importance of space assets for defence and military purpose. In the process not only are we denying ourselves a crucial capability but are obviously lagging far behind countries like China, not to talk of developed countries despite that we commenced space exploration almost about half a century back. The near total absence of literature on the subject in the country is a silent testimony to this national deficiency. Militarization of space followed by weaponisation is inevitable; hence we need to review our space policy. The question of ethics and restraint would only cripple defence preparedness. Speaking on issue the Defence analysts Raja Menon queries, "We must decide how far we are prepared to mix realism with idealism? Swinging too far towards idealism may take us back to the days of proposing an NPT and eventually to keep out of it to prevent being marginalised a non nuclear weapon state".[3]

Space presents both opportunities and dangers. The opportunities stem from the actual or potential use of space for communications, weather forecasting, remote sensing, global positioning, navigation, and many other commercial applications as well as science. The dangers stem from the possible weaponisation of space, leading to a potential arms race in space and even the accidental or intentional launching of weapons from space, which would be highly destabilising. There were decades of quiet on this front following the 1967 Outer Space Treaty, the 1972 ABM Treaty, and an implicit agreement not to conduct anti-satellite tests. Several recent developments, however, have suggested a potential for the militarisation of space with its accompanying dangers. Recent shoot-down of their own satellites by both China in January 2007 and the US in February 2008 have the potential for stimulating the race to develop anti-satellite weapons by several nations.

Space laws neither prohibit militarisation of space, nor do they prevent space vehicles from carrying conventional ordnance, but these laws

---

[3] Raja Menon in Space Security and Global cooperation 2009.

specifically forbid deployment of Nuclear weapons and Weapons of Mass Destruction (WMD). Satellites today are the basis for information passing architecture of economies and militaries. Hence, dependence on satellites today also constitutes a terrible vulnerability which is bound to be exploited by a country which feels geopolitically threatened.

India today has Satellites performing roles in almost all conceivable areas in the civil. Their use for military purposes is on the limited scale. In my view, considering that ensuing advantages outweigh the losses, the utilisation of space satellites for military use must be enhanced. However, for this we need to review our space policy. We also need to protect the space assets from possible intentional damage by the adversary who would be tempted to take advantage on our space assets which are crucial to carry out various civil and military functions. To understand Space Security from the Indian Perspective, an attempt has been made to holistically study the subject such as, what constitutes an outer space and its legal position, civil and military uses of the artificial satellites, the space policy of the major aerospace powers and the features Indian space policy. In next few chapters, after describing the India's military capability in space, and dimension of the space threat emanating from our adversaries, various options available to protect our space assets has been deliberated. Finally, the last chapter summarises the conclusions that can be drawn from the material presented and offers recommendations.

-Author

ABBREVIATIONS

ASAT	-	Anti Satellite
ADCOS	-	Advisory Committee on Space Sciences
ABM	-	Anti Ballistic Missile
ASO	-	Air-and-Space Operations
BMD	-	Ballistic Missile Defence
CPSU	-	Communist party of Soviet Union
COPUOS	-	Committee on the Peaceful Uses of Outer Space
C.I.S	-	Commonwealth of Independent States
C4ISR	-	Concept of Command, Control, Communications, Computers, Intelligence, Surveillance and Reconnaissance
CNES	-	National Space Research Centre. (implements France's Space Policy)
CD	-	UN conference on Disarmament
CNSA	-	China National Space Administration
DEW	-	Directed Energy Weapons
DOT	-	Department of Telecommunications
DOS	-	Department of Space

DRDO	-	Defence Research and Development Organization
ESA	-	European Space Agency
EU	-	European Union
EMP	-	Electro-Magnetic Pulse
FGFA	-	Fifth Generation Fighter Aircraft
FASOC	-	Future Air and Space Operational Concept (of RAF's)
GAGAN	-	GPS and GEO Augmented Navigation
GSC	-	Guyana Space Centre
GMES	-	Global Monitoring for Environment and Security
GLONASS	-	GPS and Global Navigation Satellite System of Russia
GEO	-	Geo Stationary Orbit
GPS	-	Global Positioning System
GSLV	-	Geo-stationary Satellite Launch Vehicle
HUMINT	-	Human Intelligence
HST	-	Hypersonic Transport
HSP	-	Human Spaceflight Programme
ICAO	-	International Civil Aviation Organization
ISRO	-	Indian Space Research Organisation
INSAT	-	Indian National Satellites
IRS	-	Indian Remote Sensing Satellite

IGDMP	-	Iintegrated Guided Missile Programme
IRNSS	-	Indian Regional Navigation Satellite System
IMINT	-	Imagery Intelligence
ICC	-	INSAT Coordination Committee
IISU	-	ISRO Inertial Systems
ISAC	-	ISRO Satellite Centre
ICO	-	Intermediate Circular Orbit
JDAM	-	Joint Direct Attack Munition
J.A.F	-	Joint Armed Forces of CIS
JSIC	-	Joint Space Intelligence Centre
KE ASAT	-	Kinetic Energy Anti-Satellite
LEO	-	Low Earth Orbit
LRTR	-	Long-Range Tracking Radar
LPSC	-	Liquid Propulsion Systems Centre
MEO	-	Medium Earth Orbit
MCF	-	Master Control Facility
MOM	-	Ministry of General Machine-Building in erstwhile Soviet Union
MED	-	Military Encyclopaedia Dictionary (pertains to erstwhile Soviet Union)
NCW	-	Network Centric Warfare
NNRMS	-	National Natural Resources Management Systerm

NASA   -   National Aeronautics and Space Administration

NRSA   -   National Remote Sensing Agency

NARL   -   National Atmospheric Research Laboratory

NE-SAC   -   North Eastern-Space Applications Centre

NATO   -   North Atlantic Treaty Organisation

PLA   -   People's Liberation Army

PLAAF   -   People's Liberation Army Air Force

PSLV   -   Polar Satellite Launch Vehicle

PCNNRMS   -   Planning Committee on Natural Resources
                 Management System

PRL   -   Physical Research Laboratory

PAROS   -   Prevention of Arms Race in the Outer Space

RISAT   -   Radar imaging Satellite (of India)

RKA   -   Russian Space agency

RMA   -   Revolution in Military Affairs.

SCL   -   Semi-Conductor Laboratory.

SDSC   -   Satish Dhawan Space Centre.

SAC   -   Space Applications Centre.

SME   -   Soviet Military Encyclopaedia.

SBAS   -   Space-Based Augmentation System.

SCO   -   Shanghai Cooperation Organization

SUPARCO   -   Pakistan Space and Upper Atmosphere
                 Research Commission.

SPADOC	-	Space Defence Operations Centre
SSC	-	Space Surveillance Centre
SAM	-	Surface to Air Missile
SIGINT	-	Signal Intelligence
SLV	-	Space launch vehicle
SAR	-	Synthetic Aperture Radar
TES	-	Technology Experiment Satellite (Launched by India)
UNIDIR	-	United Nation Institute of Disarmament and Research
UAV	-	Unmanned Aerial Vehicle
UNCD	-	UN Conference on Disarmament
VSSC	-	Vikram Sarabhai Space Centre
VPVO	-	Air Defence Forces (In erstwhile Soviet Union)
VPK	-	Military Industrial Commission of the U.S.S.R
WMD	-	Weapons of Mass Destruction

DEFINING OF OUTER SPACE

Introduction

What is outer space? Or, more specifically, what is the difference between national air space and outer space? The *air space* over each national territory is subject to that country's sovereign control. In *outer space*, claims of national sovereignty simply do not exist. How is one to be distinguished from the other? The question has received much attention in recent years, and many proposals on how it might be resolved have been put forward. A great deal has also been written on the subject, and several publications of the United Nations have discussed it at some length. As yet, no consensus has emerged. However, the progress of technology may make some solution more urgent in coming years. An arbitrary decision may be the only feasible answer.

Traditionally, there was no clear boundary between Earth's atmosphere and space; The Federation Aéronautique Internationale has established the Kármán line at an altitude of 100 kilometers (62 mi) as a boundary between aeronautics and astronautics. This is used because above an altitude of roughly 100 km, as Theodore von Kármán calculated, a vehicle would have to travel faster than orbital velocity in order to derive sufficient aerodynamic lift from the atmosphere to support itself. NASA's mission control, however, uses 76 miles (122 km) as their re-entry altitude, which roughly marks the boundary where atmospheric drag becomes noticeable, thus leading shuttles to switch from steering with thrusters to maneuvering with air surfaces.[4]

---

[4] http://en.wikipedia.org/wiki/Outer_space

National sovereignty over air space is a primary feature of the international agreements regarding aviation. The Convention on the Regulation of Aerial Navigation, signed in Paris on 13 October 1919, provides in Article 1 that, "every power has complete and exclusive sovereignty over the air space above its territory."[5] The basic agreement governing post-war civil aviation, namely, the Convention on International Civil Aviation, signed at Chicago on 7 December 1944, reiterates the same principle in Article 1 of the convention in virtually same language.[6]

In direct contrast, claims of exclusive national sovereignty in outer space are prohibited by international agreement. The Treaty on Principles Governing the Activities of States in the Exploration and Use of Outer Space, including the Moon and Other Celestial Bodies, was concluded in 1967 under the aegis of the United Nations. Article II provides that, *"Outer space, including the Moon and other celestial bodies, is not subject to national appropriation by claim of sovereignty, by means of use or occupation or by any other means".*[7]

The actual practice of nations also indicates a difference between national air space and outer space. Hundreds of objects have now been launched into orbit around the earth but, till date no nation has protested over their passage over its territory as violating its sovereignty. On the other hand, no nation has been willing to limit its air space to a specific height; to do so would define the upward extent of its sovereignty and, implicitly or explicitly, the lower limit of what it considered to be outer space.

---

[5] Paris Convention 1914. Convention relating to the Regulation of Aerial Navigation signed at Paris on Oct 13 , 1919 http://www.aviation.go.th/airtrans/airlaw/1914.html

[6] Convention on International Civil Aviation signed at Chicago on & Dec 1944. (Chicago convention) http://www.mcgill.ca/files/iasl/chicago1944a.pdf

[7] Treaty on principle governing the activities of the states in the exploration and use of outer space including the moon and other celestial bodies. http://www.state.gov/www/global/arms/treaties/space1.html

Schools of Thought

There are two general schools of thought regarding the need for and desirability of arriving soon at a clear line of demarcation between air space and outer space. One approach cites the need to delimit the legally binding obligations regarding the activities and authority of nations in outer space and air space, respectively. Without such a demarcation, it is contended, there will arise, as technology advances, disputes regarding the extent and nature of the obligations nations have assumed in the international agreements related to outer space. Similarly, without agreed definition, a nation could assert claims of sovereignty that would interfere with space activities desired by many other countries.

The other approach argues that there is no evidence that a demarcation line is needed and that to set one now would be premature and possibly counterproductive. The proponents of this point of view call attention to the rapid pace of space technology and the practical uncertainties regarding the characteristics of feasible and desirable space activities. Trying to set a boundary now, they feel, would risk getting it too high or defined in a way that might turn out to be detrimental to future space activities. Implicit in this viewpoint, there seems to be the expectation that the later agreement is reached, the more likely the boundary is to be set lower than it would be at present. Those who endorse a cautious approach note that the lack of specific agreement has not led to any international difficulties and does not seem likely to. They also suggest that the effort to establish a definitive boundary could, itself, lead to controversy and confusion, as has happened in regard to the demarcation between territorial waters and the high seas.

Approaches to Define the Air space & Outer Space [8]

Arbitrary Division. A layman may simply suggest setting a dividing line between air and outer space at the upper limit of the atmosphere. The

---

[8] Outer space and Air space : difficulties in definition by Dr. Raymond J. Barrett (Ph.D., Trinity College, Ireland) Department of State Advisor, John F. Kennedy Center for Military Assistance, http://www.airpower.au.af.mil/airchronicles/aureview/1973/may-jun/barrett.html

practical difficulty, however, is that the earth's atmosphere does not end abruptly; it gradually transforms into outer space. Some estimates place the altitude at which air space ceases well beyond the orbits of some existing earth satellites. In fact, there is no scientific agreement on the altitude at which air space ceases.

Characteristics of Atmosphere. A scientifically more sophisticated proposal might be to use the characteristics of the atmosphere to determine an appropriate dividing line between air and outer space. Suggestions have been made to establish the demarcation on the basis of differentiation between the several layers into which scientists divide the atmosphere. The troposphere, the layer nearest the surface of the earth, extends up to about 9 to 10½ miles at the equator and 6 to 7 miles at the poles. It is the layer in which weather phenomena occur, and it is the field of operation for conventional aviation. The troposphere contains three-fourths of all the air surrounding the earth. Most of the rest of the air in the atmosphere is contained in the next layer, called the stratosphere. It is above the weather and is reached only by the most advanced aircraft and research balloons. Its upper limit is about 25 miles. The troposphere and stratosphere contain about 99.7 percent of the air. A third layer, called the mesosphere, extends to about 50 miles, and beyond that is the ionosphere. The latter is sparsely occupied by gas particles, less dense than the most complete vacuum that can be achieved on earth. The upper limit of the ionosphere is not defined.

The major difficulty in trying to define a boundary by utilizing the characteristics of the atmosphere is the lack of uniform criteria. The physical characteristics of the atmosphere and of the various layers can be judged by a variety of criteria, such as the composition of the gases, their densities and their temperatures. These properties are not uniform at a certain altitude. They can also vary with solar activity, time of day, season, region, and other circumstances. The boundaries between the layers of the atmosphere are thus not precise, uniform in height above the earth, or constant. Thus, it is not possible, because of the variance in the properties of the atmosphere, to arrive at any other boundary between air and outer space that would be precise, uniform, and constant.

Characteristics of Flight. The layman, faced with these scientific difficulties, might suggest using the characteristics of aircraft flight to arrive at an adequate boundary. Surely, he might think, we can define the height at which aircraft can actually fly, and everything above that could be considered outer space. The Council of the International Civil Aviation Organization (ICAO) defines an aircraft as "any machine that can derive support in the atmosphere from the reactions of the air other than the reactions of the air against the earth's surface." The maximum altitude at which a machine can derive support from the reactions of the air is presently estimated at about twenty one miles by the ICAO Secretariat.

One of the most widely discussed proposals for a demarcation between air space and outer space is that it be established at the altitude where aerodynamic lift yields to centrifugal force, what is known as the "Von Kármán line." To accomplish aerial flight, weight equals aerodynamic lift plus centrifugal force. Aerodynamic lift decreases with altitude because of the decreasing density of the air. Beyond zero airlift, centrifugal force takes over. This approach also involves several difficulties that seem to preclude a uniform and constant boundary. The theoretical limit of the height of air flight may increase as the result of such developments as improved cooling techniques or more heat-resistant materials. The aero dynamical forces also vary with the character and speed of the specific object involved. Moreover, the density of the atmosphere itself is not constant but is subject to a variety of fluctuations, as already noted.

Lowest Perigee of an Orbiting Satellite. If an approach based on the characteristics of the atmosphere and aircraft is not adequate, how about tackling the problem from the other side, that of outer space? For instance, could not outer space be defined as everything beyond the lowest point (perigee) of an orbiting satellite? At a certain altitude, the earth's atmosphere is too dense for an artificial satellite to stay in orbit. The lowest perigee approach would have the advantages of being in accord with existing practices in orbiting satellites and with the attitudes of countries toward objects in earth orbit.

The perigee of a durable satellite orbit at present is about 95 to 100 miles. However, with improvements in space flight technology, orbiting with continuing rocket thrust may lower this perigee to 70-75 miles. This element of uncertainty would hardly be compatible with a definition seeking to determine national sovereignty over air space.

Loss of Gravitational Pull. One suggestion has been to set the boundary at the point where the gravitational pull of the earth ceases, this approach deriving from the idea that a nation's sovereignty need only extend to the height from which an object can be dropped on its territory. However, gravity ceases very gradually at remote heights; it is not possible to indicate an exact altitude where a boundary could be drawn based on the earth's attraction. And, even if one were feasible, it would probably be much too high; one calculation, for instance, indicates that the earth's attraction in relation to the moon is dominant up to some 205,000 miles, and much farther in relation to the sun. A further practical difficulty is that the gravitational effect of the earth depends on the escape velocity of the object, which, of course, can vary.

Intermediate Zone as a Boundary for Air Space & Outer Space. Another approach tries to overcome the difficulties in defining the outer limit of the atmosphere by proposing an intermediate zone between air space and outer space. It has been noted that there exists a buffer zone between the highest altitudes reached by balloons and aircraft on the one hand and the lowest altitude at which satellites remain in orbit without any means of propulsion on the other hand. This proposal suggests an appropriate international regime in this area, between the national sovereignty of air space and the freedom of outer space. One immediate difficulty with this approach is that the present intermediate zone is likely to narrow with technological developments and may well disappear entirely.

Moreover, the proposal still does not solve the difficulties in finding uniform and constant criteria that would make precise dividing lines between the zones possible.

Based on Country's Capability to Apply its Authority Effectively. An effort has been made to get around all these problems of a scientific definition by proposing that the exclusive sovereignty of an underlying country should extend as high as it could effectively apply its authority. This principle has often been asserted in efforts to analyze the scope and effects of the international agreements governing civil aviation. However, it has equally been challenged on the grounds it would produce unacceptable disparities, conflicts, and uncertainties. Since nations are at widely different levels of scientific and technical development, their air spaces would vary greatly. If each country were allowed to project its sovereignty upward and sideward in accord with its effective power, conflicting claims would seem highly likely to occur; and there would be no way to resolve them. The criterion of effective power would also create marked uncertainties because sovereignty would vary with the development of technology.

Based on Distinction between Aeronautical & Astronautically Activities. Another proposal suggests that a distinction be made between aeronautical and astronautical activities, rather than trying to decide on a demarcation between air space and outer space. The proponents of this approach argue that a legal definition is usually needed to permit certain activities and prohibit others. Accordingly, they feel that in regard to outer space activities, it would be better to seek this objective, not by trying to set boundaries but by defining objectives and missions for space vehicles. Their thought is that the important interests of all countries can be protected more effectively, not by putting territorial limits to national sovereignty but by legally prohibiting those actions in the course of space activities that would endanger these interests. This approach proposes that astronautical activities should be subject to one and the same legal regulation, irrespective of the altitude at which they are carried out. It would apply to them the moment they leave the earth, in order to avoid a complicated determination of their passing from one legal status to another. The only way to preserve the logical unity of legal regulation, it is contended, is by dispensing with a demarcation in space and adopting a functional criterion.

However, there are difficulties with this approach, too. It is not always possible to distinguish precisely between space activities and other activities. Using the purpose of each activity as the criterion has been suggested; but often this could be ambiguous. Moreover, the prospects of scientific and technical progress in the development of aircraft and space vehicles make the practical problem of distinguishing between them ever more complicated. Another intricate problem is how nations could differentiate between space activities at low altitudes and air activities, so as to regulate each effectively and discretely.

Conclusion

The only conclusion from a review of the various approaches to differentiating between air space and outer space is that no fully satisfactory answer is in sight. In fact, each of the approaches seems to have at least one serious defect. The problem has not been a pressing one. Indeed, the many uncertainties and potential developments in space activities have even suggested some wisdom in waiting until man's abilities and needs in space are much better defined. In the not too distant future, machines capable of flying along a ballistic trajectory are expected to orbit the earth, fly in outer space and air space, and make soft landings on the earth. The space shuttle, which NASA hopes to make a follow-on programme to the projected Apollo and Skylab series, apparently will be such a vehicle. Aeronautical researchers are reported to be thinking about a hypersonic transport (HST) as a next step after the supersonic transport, Some of the features being considered are described as "rocket-assisted take-offs" and "space vehicle-like bursts beyond the atmosphere followed by semi-orbital 'free fall' until descent."

Developments such as these are bringing closer the day when some formula will be needed, as a practical matter, to accommodate the differences between air space and outer space. The difficulties involved in all the approaches that have been suggested indicate that the decision may well have to be an arbitrary one. The goal obviously should be to select a boundary that seems to balance best the varying difficulties, advantages, and other pertinent considerations. Some demarcation line in the 50- to 75-mile altitude range may be the most satisfactory—or least unsatisfactory.

LEGAL ASPECTS OF OUTER SPACE

Introduction

Though there is no boundary line between airspace and outer space, both jurists and Governments are agreed that the sovereignty of the state does cease at some point in the space beyond which it is free. Various theories and views have been propounded. One of the theories advances the view that outer space as well as other planets and celestial bodies fall into the category of *res nullius* thereby implying that nations are free to appropriate them by traditional or other method recognized in international law. On the other extreme, is the view that space and other planets are perpetually free for the exploration and use of all nations of world. In between these two views, there are several other shades of opinion. Of which the most commendable seems to be the view that outer space and other celestial bodies should be subjected to some international control to prevent their misuse.

Three analogies have been suggested to determine the status of outer space:-

(a) A basic principle of international air law is that every state has complete and exclusive sovereignty over the airspace above its territory. One section of experts believes that the existing regime of international law governing air navigation is adequate to be applicable to the outer space. The principle of airspace sovereignty was unequivocally affirmed in the Paris Convention of 1919 and restated in the Chicago Convention on International Civil Aviation (1944)[9].

[9] Chicago convention available at http/www.mcgill.ca./files/iasl/chicago1944a.pdf

(b) The Rules of International Law applicable to the High Seas should be applied to space. It is suggested that the law of high seas offers many beneficial provisions which can be embodied with advantage in the rules of international law meant to govern outer space. The problem of trespass by spacecraft during re-entry in to the earth's atmosphere should be solved by analogy to the law of innocent passage through territorial waters in maritime law.[10]

(c) The third view is that the Legal Principles applicable to Polar Regions Antarctica are suitable for application to space. The Antarctica Treaty lays down that 'Antarctica shall be used for peaceful purposes only' and prohibits' any measures of a military nature, such as establishment of military bases and fortification, carrying out of military manoeuvres, as well as the testing of any type of weapons.[11] The similarities between outer space and Antarctica such as uninhabited by human beings, the difficulty of access, the lack of knowledge of the resources, suggest Antarctica as an appropriate analogy to the outer space.

The use of outer space is developing in two directions. On the one hand, space technology is used in every corner of contemporary human life; on the other hand, the world has experienced accelerating steps of outer space militarisation. The traditional military use of outer space has spread from supportive roles such as communication, navigation, reconnaissance, surveillance and early warning at peacetime, to direct war fighting roles such as command and control, warhead identification, target positioning and bomb guiding. Even worse, outer space is facing an urgent danger of being weaponised and becoming a battlefield. The idea of concluding a legal instrument to stop the dangerous military use is desirable but unlikely under the present circumstances with opposition from major powers.

---

[10] United Nations Convention on The Law Of The Sea-http://www.un.org/Depts/los convention_agreements/texts/unclos/closindx.htm United Nations

[11] The Antarctic Treaty-http://www.nsf.gov/od/opp/antarct/anttrty.jsp

Laws Relating to Control and Regulation of Outer Space

There are multiple players in the space debate, each having a political, economic or moral stake in the future of outer space. The inception of the field of space law began with the launch of the world's first artificial satellite by the Soviet Union in October 1957. Since that time, space law has evolved and assumed more importance as mankind has increasingly come to use and rely on space-based resources. Beginning in 1957, nations began discussing systems to ensure the peaceful use of outer space. Bilateral discussions between the United States and USSR in 1958 resulted in the presentation of issues to the UN for debate. In 1959 the UN created the Committee on the Peaceful Uses of Outer Space (COPUOS). COPUOS in turn created two subcommittees, the *Scientific and Technical Subcommittee* and the *Legal Subcommittee*. The COPUOS Legal Subcommittee has been a primary forum for discussion and negotiation of international agreements relating to outer space.[12]

International law in the field of outer space is now embodied in five treaties and five sets of legal principles evolved by the United Nations.[13] These essentially establish the principle of freedom of exploration and use of outer space by all nations, declaring that outer space shall be the province of all mankind. Accordingly, outer space including the moon and other celestial bodies are not allowed to be subject to national appropriation. Placement or use of nuclear weapons and any other kinds of weapons of mass destruction in outer space are prohibited. 'States' have to bear international responsibility for their activities in outer space, whether such activities are carried out by the government entities or non-government entities under their jurisdiction. These treaties *inter-alia*, also deal with aspects like liability for damage caused by space objects, the safety and

---

[12] http://www.oosa.unvienna.org/oosa/COPUOS/copuos.html

[13] http;//www.britannica.com/EBchecked/topic/10733/air-law/39247/ Airspace#ref=ref424625, United Nations Treaties and Principles on Outer Space United Nations Treaties And Principles On Outer Space. (www/oosa.Unvienna.org/oosa/spacelaw/treaties/html)

rescue of spacecraft and astronauts, prevention of harmful interference for the space activities, avoidance of adverse changes in the earth's environment, prevention of harmful contamination of celestial bodies, obligations for notifying/registering objects launched into outer space and procedures for settlements of disputes. Further, the five declarations and sets of legal principles, adopted by the General Assembly, in the nature of 'soft laws' address the aspects related to promotion of international cooperation in space activities, the dissemination and exchange of information through international direct television broadcast via satellite, sharing of data and information from satellite earth observations and guidelines for safe use of nuclear power sources in outer space.

International Treaties and Agreements [14]

Five international treaties have been negotiated and drafted in the COPUOS:

(a) **The Outer Space Treaty, 1967.** This Treaty lays down the principles for governing the activities of states in exploration and use of the Outer Space, including the Moon and Other Celestial Bodies. It bars States from placing nuclear weapons or any other weapons of mass destruction in orbit of Earth, installing them on the Moon or any other celestial body, or to otherwise station them in outer space. It exclusively limits the use of the Moon and other celestial bodies to peaceful purposes and expressly prohibits their use for testing weapons of any kind, conducting military manoeuvres, or establishing military bases, installations, and fortifications (Art.IV). However, the treaty does not prohibit the placement of conventional weapons in orbit. The treaty also states that the exploration of outer space shall be done to benefit all countries and shall be free for exploration and use by all States. (The "Outer Space Treaty", adopted by the General Assembly in

---

[14] ibid

its resolution 2222 (XXI), opened for signature on 27 January 1967 and entered into force on 10 October 1967).

(b) **The Rescue Agreement, 1968.** This agreement calls for the rendering of all possible assistance to astronauts in the event of accident, distress or emergency landing, the prompt and safe return of astronauts, and the return of objects launched into outer space (The "Rescue Agreement", adopted by the General Assembly in its resolution 2345 (XXII)), opened for signature on 22 April 1968 and entered into force on 3 December 1968);

(c) **The Liability Convention, 1972.** The convention lays down the State and International Liability for Damage Caused by Space Objects.(The "Liability Convention", adopted by the General Assembly in its resolution 2777 (XXVI)), opened for signature on 29 March 1972 and entered into force on 1 September 1972);

(d) **The Registration Convention, 1975**. The convention lays down that each satellite launching state shall maintain a State Registry which shall furnish to the Secretary-General of the United Nations, as soon as practicable, the relevant information concerning each satellite launched in the space. The Secretary-General of the United Nations shall maintain a register in which the information furnished by the satellite launching state shall be recorded. Similarly, each state registry shall also notify the Secretary-General as soon as practicable the information of space objects which it has previously transmitted information, and which are no longer in Earth's orbit. (The "Registration Convention", adopted by the General Assembly in its resolution 3235 (XXIX)), opened for signature on 14 January 1975 and entered into force on 15 September 1976);

(e) **The Moon Treaty, 1979**. To prevent the Moon and Other Celestial Bodies, from becoming an area of international conflict, the treaty stipulates that the Moon and other celestial bodies shall be used by all States Parties exclusively for peaceful purposes. Any threat or use of force or any other hostile act or threat of hostile acts is prohibited. It is likewise prohibited to use the Moon in order to commit any such act or to engage in any such threat in relation to the Earth, the Moon, spacecraft, and the personnel of spacecraft or manmade space objects. It also lays down that states shall not place in orbit around or other trajectory to or around the Moon objects carrying nuclear weapons or any other kinds of weapons of mass destruction or place or use such weapons on or in the Moon. (The "Moon Agreement", adopted by the General Assembly in its resolution 34/68), opened for signature on 18 December 1979 and entered into force on 11 July 1984).

Legal Principles. The five sets of legal principles adopted by the United Nations General Assembly provide for the application of international law and promotion of international cooperation and understanding in space activities are given below:-

(a) **The Declaration of Legal Principles Governing the Activities of States in the Exploration and Uses of Outer Space**. These principles primarily stipulate that exploration and use of outer space shall be carried on for the benefit and in the interests of all mankind. Outer space and celestial bodies are free for exploration and use by all states on a basis of equality and in accordance with international law. These are not subject to national appropriation by claim of sovereignty, by means of use or occupation, or by any other means. The activities of states in the exploration and use of outer space shall be carried on in accordance with international law (General Assembly resolution 1962 (XVIII) of 13 December 1963);

(b) **The Principles Governing the Use by States of Artificial Earth Satellites for International Direct Television**

Broadcasting. Every state has an equal right to conduct activities in the field of international direct television broadcasting by satellite and to authorise such activities by persons and entities under its jurisdiction. All states and peoples are entitled to and should enjoy the benefits from such activities. Access to the technology in this field should be available to all States without discrimination on terms mutually agreed by all concerned (Resolution 37/92 of 10 December 1982);

(c) **The Principles Relating to Remote Sensing of the Earth from Outer Space.** Remote sensing activities shall be carried out for the benefit and in the interests of all countries, irrespective of their degree of economic, social or scientific and technological development, and taking into particular consideration the needs of the developing countries (Resolution 41/65 of 3 December 1986);

(d) **The Principles Relevant to the Use of Nuclear Power Sources in Outer Space.** Activities involving the use of nuclear power sources in outer space shall be carried out in accordance with international law, and in a manner to minimize the quantity of radioactive material in space and the risks involved, the use of nuclear power sources in outer space shall be restricted to those space missions which cannot be operated by non-nuclear energy sources in a reasonable way (Resolution 47/68 of 14 December 1992);

(e) **The Declaration on International Cooperation in the Exploration and Use of Outer Space.** States are free to determine all aspects of their participation in international cooperation in the exploration and use of outer space on an equitable and mutually acceptable basis while taking into account the needs of developing countries (Resolution 51/122 of 13 December 1996).

The COPUOS operates on the basis of consensus, *i.e.* all committee and subcommittee delegates must agree on treaty language before it can be included in the final version of a treaty, and the committees cannot place new items on their agendas unless all member nations agree. One reason that the U.N. space treaties lack definitions and are unclear in other respects, is because it is easier to achieve consensus when language and terms are vague. In recent years, the COPUOS Legal Subcommittee has been unable to achieve consensus on discussion of a new comprehensive space agreement, and it is also unlikely that the Subcommittee will be able to agree to amend the Outer Space Treaty in the foreseeable future. Many space faring nations seem to believe that discussing a new space agreement or amendment of the Outer Space Treaty would be futile and time consuming, because entrenched differences regarding resource appropriation, property rights and other issues relating to commercial activity make consensus unlikely.

Outstanding Issues at International Level

Beginning with the 'Moon Agreement', which came into force on 11 July 1984, the progress in development of space law has been slow. First, the Moon agreement itself received very limited response of international community with just 13 ratifications and 4 signatures (as of 1st January 2008) Major space faring nations such as the USA and Russia did not sign this agreement. The principle of' Common Heritage of Mankind', which was basically derived from the Laws of Sea, is the bone of contention, which has thwarted the wider acceptance of Moon agreement. Ironically, the major space treaties were made in the then prevailing atmosphere of cold war confrontation between the USA and the erstwhile Soviet Union. In the subsequent years, the context of actors in space activities has changed considerably. More countries are now actively engaged in developments and applications related to space technology besides, participation by private sector in these activities. Even as the space technology and its applications advanced rapidly, the tardy further development of international legal framework has left many questions unresolved, prominent issues such as definition of outer space and the demarcation of boundary between outer space and airspace, which is governed by different legal regimes, has not been agreed upon. The definition of the term 'peaceful uses of outer space' itself is subject to varied interpretations. While some argue for totally pacific applications of space, other consider that there is scope for military uses of space including testing or using weapons in space, other than placement or use of nuclear weapons or weapons of mass destruction. Notwithstanding the fundamental nature of unresolved issues, these did not impede the progress of space activities. However, if space activities rapidly expand further, many issues for law are to be resolved with greater speed.

Space Debris. Another issue of concern is that of Space Debris. Although debris in space is growing with increasing potential for collisions, there is no agreement for evolving legal measures and the UN Committee on Peaceful Uses of Outer Space could adopt a set of voluntary guidelines after considerable debate. As of today, space agencies have been adopting certain voluntary measures.

Equitable Access and Use of Geostationary Orbit (GSO). The other issue is that of ensuring rational and equitable access and use of Geostationary Orbit (GSO), which is commonly used by communication Satellites. Over the years, GSO is crowded with satellites and new entrants find it difficult to get access to orbit/ spectrum resources. There is a high level of private sector activity in this domain and issues relevant to prevention of market dominance, technology transfer, public service objectives and effective enforcement mechanism need further regulatory provisions.

Private Sector Participation. The third issue relates to private sector participation in commercial exploitation of extraterrestrial resources. In the past, private industries were actively involved in activities, such as satellite communications, remote sensing and provision of launching equipment and services. Now, private industry is also showing interest to participate in activities such as space tourism, mining of asteroids and even, waste disposal in outer space. However, given the emphasis accorded in the international space law to the state responsibilities for all activities in outer space, private sector finds that the current legal framework is inadequate for growing their activities and role.

Mechanism for Settlements of Disputes. There are also demands for establishing effective mechanism for the settlement of disputes, arising in relation to space commercialisation.

Restricted and Regulated Flow of Technologies. Under the International Space Law, in particular the Outer Space Treaty, one observe that at the very outset, a statement under Article-1 that the exploration and use of outer space including the moon another celestial bodies, shall be carried out for the benefit and in the interests of all countries, irrespective of their degree of economic and scientific development, and shall be province of all mankind. However, even after four decades of existence of the treaty, the capacities and level of access to outer space by different countries vary widely, across the globe. Flow of technologies relating to space systems and even some of their applications are considerably restricted and regulated based on national legislations or multilateral export control regimes. This situation partly is triggered by the concerns on the dual use nature of space technology. Achievement of equitable access to space by all countries is still remains an unfulfilled need, which the current international legal system is unable to redress. A further lacuna is the absence of a common agreement on ethics and an enforcement mechanism.

Notwithstanding the above limitations or slower pace of progress in recent times, it is highly noteworthy that the international space law as it evolved, set a unique trend and standard in establishing the freedom of access and preventing national appropriation of space which is considered vital for connecting global community. In addition, space provided unique tools for assessing and monitoring the health of Planet Earth itself.

National Law [15]

It is borne out by experience that national legislations in the areas of growing importance to economic and social sectors of the nation can lead to an orderly development that enhances benefits. Several countries such as USA, Russia, Australia, Canada, UK, Sweden and Israel have enacted national space legislations to regulate and guide their space activities. A need, which is being debated currently in case of India too. The need for national space legislation has been examined from the viewpoint of harmonising the domestic legal environment with the specific obligations arising from international treaties related to space, in which India is a party. The national space legislation, in particular, should address the specific interests and needs of Indian society.

Space and space-related matters in India are regulated by legal rules belonging to different areas of the Indian domestic law, since there is no special space legislation. The legal position of the space industry is largely determined by the Constitution of India as the constitutional provisions relating to general international law are also relevant to the aerospace law.

Article 51 of the Constitution imposes on the state the obligation to strive for the promotion of international peace and security, including maintaining just and honourable relations between nations, respect for international law and treaty obligations, and settlement of international disputes by arbitration.

Under Article 73 the executive power of the Union extends (a) to the matters with respect to which Parliament has power to make laws and (b) to the exercise of such rights, authority and jurisdiction as are exercisable by the government of India by virtue of any treaty or agreement.

Article 245 empowers Parliament and state legislatures to enact laws. The Constitution enumerates three lists of subjects - the Union list, the

---

[15] An article on *"Legal Environment for Space activities"* by K R Sridhara Murthy, V Gopalakrishanan and Parthasartha Sarthi Datta of Antrix Corporation Ltd.

State list and the Concurrent list - in respect of which the legislative power may be exercised, provided that the legislative power of Parliament overrides that of the state legislature in respect of the concurrent list. The "space" as a subject is not mentioned in the Union or Concurrent List The reason was that the Constitution was adopted in 1950, but the space activities started in India in the early Sixties; a number of items on the Union list related to the aerospace activities in India.

Article 248, Parliament retains residuary legislative power in respect of "any matter not enumerated" in any of the three lists.

Article 253 empowers Parliament to make any law for the whole or any part of the territory of India for implementing treaties, international agreements and conventions. It enables the Government of India to implement all international obligations and commitments. Following the commonwealth practice treaties are not required to be ratified by Parliament in India. They are, however, not self-executory. Parliamentary legislation is necessary for implementing the provisions of a treaty within the country. Parliament has passed many Acts to implement international treaties and conventions (including environment, civil aviation etc.) but not in outer space activities.

Conclusion

India is a party to all important space treaties, which form the main body of international space law i.e., the Space Treaty 1967, the Rescue Agreement of 1968, the Liability Convention of 1972, the Registration Convention of 1975, and the Moon Agreement of 1979.However, we are notoriously slow, in acceding to treaties and incorporating its convention obligations in municipal law. This has resulted in the absence of any legislation relating to space activities in India.

Up till now, there is no comprehensive or specific law dealing with space activities in India. However, with the rapid development of activities in space, there is a growing need for enacting a new domestic space law, and integrating divergent regulations dealing with space and space-related matters. Such a law should define the role of the DoS and its various organs and different governmental and non-governmental agencies in space matters, the procedure for adoption and implementation of space programmes, and regulations on the safety of launch and space flight, the question of transit of foreign space objects through national airspace, questions of liability and insurance, protection of intellectual property rights, spin-off benefits, and above all, implementation of international obligations under the various treaties. Further, it should also formally incorporate the objectives of India's space policy, reiterating the country's commitment to the peaceful uses of outer space.

SPACE POLICIES OF MAJOR SPACE POWERS

Presently, amongst the existing ten space faring nations, three major players are seriously pursuing programmes to enhance their capabilities in space warfare. United States, erstwhile soviet union (now Russia) were in the race since early sixties. The China, the last entrant in the race is also actively working to gain offensive and defensive capabilities in the space. European Space agency, Japan and India pursue civilian programme. Unlike India, countries of the European Union and Japan however have the protection of US defence umbrella. United Kingdom and France don't want to remain under the shadow of United States hence, follow independent space programmes to bolster their military capabilities. Israel too has launched several spy satellites to keep tab on Iran possessing credible missile capacity to strike Israel. Its apprehension has further increased with Iran's quest for nuclear capability. At present, only eight to 10 countries routinely apply space capabilities to their military operations with India joining them soon with its first dedicated military satellite to support command and control functions of Indian Navy[16]. There are another 45 or so nations (Iran being the 45th nation) with civil space capabilities that could eventually be applied to military uses. As only handful of space faring nations have the capability to launch satellites. Other nations pay them for launching their satellites in the space.

US Space Policy

Satellites offer an excellent mean of monitoring and surveillance of the

---

[16] Rajat Pandit in an article "*Dedicated satellite for the navy by the year end*" Times of India 20 May 2010

adversary; hence, from the beginning of the space age, U.S. always sought
the outer space to be recognized as free territory not subject to the normal
confines of territorial limits of any state. In early sixties, United States
propagated the concept of "freedom of space" but, the Soviets suspecting
American motives were opposed to this. Soviet Union's opposition,
however, changed on 4 October 1957 after the launch of Sputnik which
also ushered in the space age. The satellite flew over a multitude of nations,
including the United States, without provoking a single diplomatic protest
thus establishing the precedent of "freedom of space.[17]

Formal U.S. space policy was first enunciated in the presidential term
of Jimmy Carter (Administration's PD/NSC-37 of 1978). The top goals
of the policy were to "strengthen the nation's space leadership and ensure
that space capabilities are available in time to further U.S. national security,
homeland security, and foreign policy objectives" and to "enable unhindered
U.S. operations in and through space to defend own interest there".[18]
The policy has changed little since then. On Aug 2006, United States
issued a new U.S. National Space Policy. The policy specifies that space
capabilities including the ground and space segments and supporting links
vital to US national interest. Accordingly, the United States aims to preserve
its rights, capabilities, and freedom of action in space and dissuade or
deter others from either impeding those rights or developing capabilities
with intention to do so.[19] The directive therefore, calls for permanent US
dominance in space, including the right to deny any nation access to space
if its actions are seen as hostile. American strategists believe that:-

(a) **Space is a Place.** It is a geographic location like air, land, and
sea as such America's military capabilities must include the ability
to control the use of space and defend its military and civil assets

---

[17] The Freedom of Space doctrine, www. American foreign relations .com/o-w/outer space

[18] Presidential Directive NSC-37 US National Space Policy, 11 May 1978, available at http:/
/www.au.af.mil/au/awc/awcgate/nsc-37.htm

[19] US National Space Policy, 31 Aug 2006, available at http://www.whitehouse.gov/sites/
default/files/microsites/ostp/national-space-policy-2006.pdf

from foreign attack.

(b) **A Forward Presence in Space it Vital, even during Times of Peace.** Space soon is becoming the seas of the future. That means, the U.S. must be free to navigate in space, have and protect space lanes of communication, and monitor all vessels that travel through space.

(c) **Space must be Dominated during Wartime.** This means the U.S. must be prepared to protect U.S. access to space while denying its enemies' access to space. It also means that the U.S. must be capable of exploiting all vital space regions like preferred orbits and missile lanes.

(d) **The U.S. Military must Increase America's Capabilities of Utilizing Space for a Variety of Missions.** These include: command, control, communications, computers and intelligence (C4I), tactical warning, weather forecasting, navigation, anti-satellite, space to ground attack and missile defence.[20]

US Space Command Vision 2020

US Space policy further gets reflected in US Space Command Vision 2020 in which it has talked of prominent role of space in giving Information superiority to attain full spectrum dominance while denying an adversary's ability to fully leverage the same. To move towards the attainment of vision, US Space command has adopted four operational concepts:-

(a) Control of Space.

(b) Global Engagement.

(c) Full Force Integration

(d) Global Partnerships.

---

[20] The Military Use of Space by Bryan Johnson http://www.suite101.com/article.cfm/political_economy/19993#ixzz0oYl3SjyS

These operational concepts provide the conceptual framework to transform their vision in to capabilities. For example in **Control of space** US wants to assure itself freedom of operations within the space, security to US military, civil, and commercial assets in space and an ability to deny others the use of space, if required. The concept of *Global Engagement* relates to the application of precision force from, to, and through space. The third concept of *Full Force Integration* aims for the integration of space forces and space-derived information with land, sea, and air forces and their information. Finally, the *Global Partnerships* aims to augment military space capabilities through the leveraging of civil, commercial, and international space systems. The growth of US non military space systems provides the opportunity to the United States to gain increased battle space awareness and information in a cost-effective manner by sharing the costs, risks, and increased opportunities.[21]

Today, the U.S. has a commanding lead in space technology and military space capabilities. There are around 300 military satellites and 600 commercial dual use satellites in the space. Of these, US operates nearly half of the military and spends nearly $20 billions every year on military space activities. Constellations of both military and civilian satellites provide protection and support for military operations, deliver ballistic missile early warning, provide reliable, secure and jam-proof communications, gather audio-visual and electronic intelligence, predict weather patterns, guide navigation, target weapons, and perform a host of other missions. These systems are critical to America's status as a world power, however, there are signs now that foreign powers are either on the verge of challenging America's lead or are determined to be a future competitor. America is so much concerned of the safety of their space assets that in January 2001, a commission led by Former Defence Secretary Donald Rumsfeld, warned of the growing threat to US space assets from so-called "rogue states," who might build and deploy "space mines",

---

[21] US Space Command Vision for 2020

launch ballistic anti-satellite weapons, deliberately increase orbital debris or detonate high-altitude nuclear explosions. He warned of "space pearl harbour" unless United States dominates the space. [22] Faced with the perceived challenges, America calls for "full spectrum dominance". Also, the National Security Strategy, issued by the White House in 2010, acknowledges the need to develop military capabilities across all domains –land, air, sea, space and cyber.[23]

The Air Force Space Command's most recent long-term planning document, "Strategic Master Plan FY 04 and Beyond" published in November 2002, discusses "defensive counter space" operations and "offensive counter space" operations and by 2016-2028 timeframe, the plan envisions the availability of "space-based counter space" systems, which in generic terms are known as Anti-Satellite Weapons (ASATs). The plan elaborates the American intention to field "full-spectrum, space-based Offensive Counter Space(OCS) systems capable of preventing unauthorized use of friendly space services and negating adversarial space capabilities from LEO to GEO altitudes".[24]

The US space policy leaves no doubt that it would use medium of space to further its offensive and defensive means to attain its objectives. It would also not accept any restrictive international agreements / treaties that impinge its space policy. The militarisation of space has already taken place its weaponisation is the Fait Accompli.

Presently, international community is concerned over the following developments:-

(a) US withdrawal from the ABM treaty, which removed the long-

---

[22] Background Paper: Peaceful And Military Uses Of Outer Space: Law And Policy Prepared By Institute Of Air And Space Law, Faculty Of Law, Mcgill University, Montreal, Canada February 2005.

[23] US space command vision for 2020.

[24] U.S. Air Force Space Command, "Strategic Master Plan FY04 and Beyond," pp. 12-13, http://www.peterson.af.mil/hqafspc/library/AFSPCPAOffice/Final%2004%20SMP—Signed!.pdf

standing prohibition of deploying space-based missiles.

(b) Adoption of a new *US National Space Policy* in 2006 that states that capabilities be maintained to execute space control and assert the right to deny any nation access to space if its actions are "perceived" to be hostile.

(c) Development by the US (and other space-faring nations) of weapons intended to attack targets in space.

(d) Development of technologies that have dual capabilities of protecting and attacking satellites.

Russian Space Policy

Background

After demise of the Erstwhile Soviet Union, Russia being major partner inherited the Soviet Space programme which began with ICBM development in the 1950s. In the cold war environment, the Soviet Union developed a powerful rocket and space industry, versatile R&D facilities, and an extensive infrastructure to support both missile testing and space operations. The operational support infrastructure included a number of rocket test ranges and a network of tracking, telemetry and control stations spread across the U.S.S.R.

Space Programme Management. In the Soviet Union, the Central Committee of the CPSU (with the Politburo at its head) was the supreme body, governing all military and space activities in the U.S.S.R. The responsibility for implementation of all space projects, as well as for missile programmes rested with the so-called Ministry of General Machine-Building (MOM). It was one of nine military industrial ministries supervised by the Military Industrial Commission of the U.S.S.R. Council of Ministers (VPK) and the Defense Division of the Central Committee of the CPSU. Unlike the U.S. Air Force, the Soviet Air Force had only marginal involvement in space activities. The Air Force was mainly responsible for cosmonaut selection and training and for search and rescue operations. Other military branches acted as end users of operational space systems. The exception was the Air Defense Forces (VPVO), which controlled the early warning satellite system and the anti-satellite system. VPVO had their own network of control and tracking stations for this purpose.

By the end of 1991, the Soviet Union was dissolved and all the Soviet space properties were divided among the republics. After creation of the Commonwealth of Independent States (C.I.S.), the space programme of the U.S.S.R. was replaced by a joint effort of member states, much like the European Space Agency. However, sharing the armed forces and military assets of the U.S.S.R. was a matter of constant dispute for the C.I.S. members. In December 1991, the Space Forces of the former U.S.S.R. were initially included in the Joint Armed Forces (J.A.F.) of the C.I.S. and given the global mission of joint space activities. However, disputes about the composition of the JAF continued throughout 1992, mostly because of disagreements. Consequently, Russia established its own Ministry of Defence on 16 March 1992 and numbers of units, initially to be included in the JAF, were transferred to Russian authority.

In August 1992, the Directorate of Space Means was formally incorporated into the Russian Ministry of Defense and became the Space Forces of the Russian Federation. This last step concluded the process of concentrating the Soviet space programme under the auspices of Russia. A substantial portion of the space-related industrial potential and support infrastructure remains outside Russia, particularly in Ukraine and Kazakhstan. However, the inherited space-related capabilities in Kazakhstan and Ukraine (and lesser facilities in other republics) are not comprehensive enough to enable the former republics to continue any part of the Soviet space activities or to build independent space programs, at least for the time being. According to different estimates, Russia possesses 75 percent of "space related properties" (measured by their economic value) and 90 percent of the space industry enterprises. Russia, therefore, has become the sole successor of the Soviet Space Programme.

As soon as the U.S.S.R. was dissolved, the new Russian Ministry of Industries was established to take over responsibility for all industrial enterprises situated in Russia. Within this ministry, the Department of General Machine-Building took over the enterprises of the former Ministry of General Machine-Building. The supervising ministry no longer dictates to industry exactly what they must develop. Enterprises have to bid for contracts from the Ministry of Defence. The Ministry of Defence receives its own budget for weapons procurement (including military satellites and space launch vehicles). Another major change in space took place in February 1992, when President Yeltsin established the Russian Space Agency (RKA) for implementing national space policy and developing space systems for scientific and civil applications. In so doing, responsibilities for civil and military space activity were split between the RKA and the Ministry of Defence, which gave the civilian space programme a chance to develop independently from the military programme. Throughout the 1980s the pace of Soviet space launches had slowly diminished from roughly 100 to 90 per year. A sharper drop occurred in 1989-1990, and in 1991 (the year the U.S.S.R. broke up) the launch rate dropped to 59 successful launches, the lowest in 25 years. However, this decline does not reflect any change in commitment to space activity, but was caused by economic problems, and later aggravated by political turmoil. However, as soon as budgeting and management problems were resolved, the decline in space activity stopped. At the end of 1992, the launch pace returned to the level of the late 1980.[25]

Like the United States, Russia maintains a full spectrum of military-dedicated satellite systems for early warning, communications, navigation, optical reconnaissance and signals intelligence and holds a place as a second most powerful Space power. The Russian space industry is large,

---

[25] Soviet Military Space Doctrine –www.fas.org/irp/dia/sovmilspace (Report produced by US government)

well entrenched and technically capable.[26] However, the Russian military space programme was earlier starved of funding which improved with higher earnings from exports of oil and gas resource from 2005 onwards.[27]

Russian Military Space Doctrine

Russian Military space doctrine is essentially follows the erstwhile Soviet military space doctrine, which functioned under the Soviet military doctrine. Likewise, the two key elements of the Russian Military Doctrine are the overwhelming offensive application of the superior military force to further Russian interests and the combined approach to the combat operations. The other aspect of Russian military doctrine is uniform application of military concepts to the armed forces on the whole and to each of the five military services which included Strategic Rocket Forces and Air Defence Forces, the two Russian Armed Forces involved in Russian space programme. Even in recent Russian Federation Military Doctrine (2010), the space continues to be one of the constituent of the overall military strategy.[28]

The writing of Marshal VD Sokolovskiy, former Chief of General Staff of the Soviet Forces had a seminal influence on the Soviet Space Doctrine. He claimed that "considerable part of the US programme on mastery of the space for the military purposes is for creation of the anti space weapons for the destruction of the Aerospace vehicles. He asserted that it would be a mistake to allow the imperialist camp to achieve superiority in this field and we must oppose the imperialists with more effective means and methods for the use of space for defence purposes. Only this way we can force them to renounce the use of the space for the destructive and devastating war".[29] The 1968 version of the Soviet Military strategy

---

[26] Developments in Military Space: Movement toward space weapons?By: Theresa Hitchens

[27] *en.wikipedia.org/wiki/Russian Federal_Space_Agency.*

[28] The Military Doctrine of the Russian Federation, http://www.carnegieendowment.org/files/2010russia_military_doctrine.pdf

[29] Soviet Military Space Doctrine –www.fas.org/irp/dia/sovmilspace (Report produced by US government)

outlined the Soviet's views on the military use of space which was based on the following three paths which were broadly analogues to the American doctrine i.e., Force enhancement, space control and force application[30]:-

(a) Creation of space weapon systems to assure combat effectiveness for all branches of the armed forces.

(b) Preventing others from utilizing space.

(c) Development of the strategic Offensive Systems to conduct battles in the space.

The soviet Military leadership began formulating a military doctrine at some point in the 1950s. The Sokolovskiy writing (Soviet Military Strategy 1962) provides insight of soviet space doctrine. He writes about "the problems of using outer space for military purposes", "the mastery of the space" orbital bombers, anti satellites etc. In fact, Soviet Union carried out the first anti satellite test in 1968 which showed extent of their capability. Other documents that give hints of the Soviet Space Doctrine was Soviet Military Encyclopaedia (SME) and Military Encyclopaedia Dictionary (MED) wherein Terms like Supremacy in the Space, Space weapons, Selection of the Aerospace targets, Means of detection of Aerospace Targets, Aero space situations etc are discussed. Such developments have been ascribed to other countries; especially to the United States while at the same time not admitting own capability. These observations however, were indicative of Soviet Strategy. In all Soviet writings, Soviet interest in the outer space is has always been framed in terms of defence against the enemy space attacks. It showed Soviet intention to explore the possibility of defence from the space borne attack and by extension, deny the enemy the opportunity to gain supremacy in the space. In its view such an objective is best accomplished by Soviet acquisition of such supremacy.[31] It needs to be appreciated that Soviet's did not consider

---

[30] *Space ,The frontier of modern defence,* 2006, by Squadron Leader KK Nair.

[31] Soviet Military Space Doctrine –www.fas.org/irp/dia/sovmilspace (Report produced by US government)

offence and defence as mutually exclusive or even opposing concepts. Ideologically any weapon developed by them by definition is defensive, while soviet consider offence as the most basic component of military operations, a defensive component is essential part of overall strategy. Strategic offensive and defensive systems must work synergistically to achieve victory. This is best exemplified by the soviet interpretation of the term "Air Defence". To the Soviets, an air defence operation is an anti air operation which has the intention of leading to the air supremacy. Thus an inherently defensive operation becomes offensive in nature and vice versa.[32]

In this sense, Russian 2010 military doctrine defines itself as strictly defensive. The doctrine points out 11 actions which constitute "external dangers to the Russian Federation and includes NATO and moving of the NATO infrastructure closer to its borders as threat. With regards to space, it explains that attacks on its early warning and space surveillance systems would represent a direct threat to its security. Therefore, a basic Russian national security objective is the protection of Russian space systems. The space systems also include ground stations on its territory. During peace time the main task of the armed forces is to provide timely warning to the Russian Federation Commander of an air and space attack and to maintain readiness to launch an air and space attack on the enemy to counter any threat besides, maintaining the orbital space devices to support the activities of the Russian Federation Armed Forces. Latter concern derives from Russia's assessment that modern warfare is becoming increasingly dependent on space-based force enhancement capabilities. During war, the main task of the Armed Forces is to inflict defeat on the aggressor, force him to cease the hostilities on the terms that meet the interest of the Russian Federation.[33]

Russia has expressed concern about the potential weaponisation of space and the extension of the arms race to outer space, especially in light

---

[32] Space ,The frontier of modern defence 2006, by Squadron Leader KK Nair.
[33] http://www.carnegieendowment.org/files/2010russia_military_doctrine.pdf

of the development of US missile defence systems. Russia has actively argued for a treaty prohibiting the deployment of weapons in space. In the interim, Russia has pledged not to be the first to deploy any weapons in outer space and has encouraged other space-faring nations to do the same. However, various Russian officials have also threatened retaliatory measures to any country that attempts to deploy weapons in space.[34]

As per one view, at present the military support functions are more important to Russian Federation but, as its capability of conducting the warfare in space becomes less technologically constrained to the Russian Federation, this function will gain more importance vis-à-vis the support functions. In any case the ability to provide space based military support for terrestrial combat support requires freedom of operation, if not, outright dominance of, outer space. Russia fully recognizes the decisive need to disrupt, if not destroy, enemy command, control and communication assets. It is more likely that Russia would turn to a policy of "asymmetric response," to counter the systems developed by the United States should they present a threat to Russia's space assets. In the prevailing political environment of Russia, efforts are also being made to shift the priorities in the national space programme from military to civil applications.[35]

Chinese Space Policy

China believes that US poses a "significant", long-term challenge to the country and appears to be strengthening its war fighting skills with an armed forces modernization programme. From 1990s, the PLA began to vigorously promote RMA with Chinese characteristics. It began to adopt a strategy of strengthening the military by means of science and technology. It has set for itself a strategic goal as defined by it as building informationised military and winning informationised wars.[36]

---

[34] http://www.spacedebate.org/evidence/2211/

[35] ibid.

[36] White Paper on China's National Defence 2009, www.china.org.cn/government/central_government/2009-01/20/content_17155577_4.htm

China's public posture, however, is that Outer Space should be used exclusively for peaceful purposes. It is opposed to any militarization of space. In its 1998 and 2000 white papers on national defence, it called for the creation of a multilateral mechanism to prevent an arms race in outer space. Ironically, on 11 Jan 2007, China, itself destroyed an old Chinese weather atellite in an Anti Satellite test which bellied its tall claims about peaceful use of outer space. Some analysts therefore, believe that, China is covertly developing military space assets, including anti-satellite technology. This view is supported by writings of various China's military analysts, as well as media which have highlighted the importance of the development of a military space program to defend against U.S. dominance in the future.[37]

China's Counter Space Strategy. Development of military space capabilities is the key element in the Chinese armed forces modernization program. In this, creation of anti-satellite weaponry, building new classes of boosters, as well as improving an array of military space systems are of pivotal importance.[38] The development and testing of ground based anti satellite weapon indicates that china in near terms is moving towards a doctrine of deterrence with counter space capability. The doctrine proclaims an officially preferred ban on all space weapons; deterrence based on finite capability rather than on competition with United States with preference to ground based weapons than the space based weapons. In Feb 2008, China joined Russia and proposed a treaty to prevent weaponisation of the outer space at the UN Conference on Disarmament (UNCD).[39] However, Chinese writings make it clear that in long-term; it envisions conflict in the space and is preparing for it.

According to the Chinese, the strategic competition in the 21st century

---

[37] China's Attitude toward Outer Space Weapons. http://www.nti.org/db/china/spacepos.htm

[38] Background Paper "Peaceful" And Military Uses Of Outer Space: Law And Policy Prepared By Institute Of Air And Space Law, Faculty Of Law, McGill University, Montreal, Canada February 2005

[39] China , Space Weapons and US Security http://www.cfr.org/publications/16707/

will not be on Earth, but in space. Towards this objective, for more than a decade, Chinese military strategists and aerospace scientists have been quietly designing a blue print for achieving space dominance[40]. This view gets credence with the statement General Xu Qiliang, the present commander of the People's Liberation Army Air Force, argues that space exploration is critical to China's national security interests. His remarks reflect the Chinese government's growing interest in space exploration and the development of space technology.[41]

The Chinese regard outer space as the fifth-dimension operational space after land, sea, air, and electromagnetism. It believes in developing the "new-concept" space weapons, viz: laser weapons, ultra-high frequency weapons, ultrasonic wave weapons, stealth weapons, mirror-beam weapons, electromagnetic guns, plasma weapons, ecological weapons, logic weapons, and sonic weapons. Because the space theatre of war is in outer space, there are no restrictions of national boundaries and sovereign air space. The side possessing space dominance can therefore exercise complete freedom of action.[42]

Space Warfare

Chinese military scientists contend that space warfare will become the core of future non-contact combat. The integrated space-based combat platforms, weaponry, and C4ISR components will guide the various combat elements of the three armed services to launch long-distance precision attacks on ground, sea, air, and space targets. With the continuing development of space weaponry and equipment, belligerents will conduct various modes of space warfare such as:-

   (a) Space information warfare,

---

[40] *China's Military Strategy in Space* by Mary c. Fitzgerald. http://www.hudson.org/files/publications/07_03_29_30_fitzgerald_statement.pdf

[41] http://csis.org/blog/chinese-space-policy-collaboration-or-competition

[42] *China's Military Strategy in Space* by Mary c. Fitzgerald, Research Fellow, Hudson Institute available at www.hudson.org/files/publications/07_03_29_30_fitzgerald-statement.pdf

(b) Space electronic warfare,

(c) Space anti-satellite warfare,

(d) Space anti-missile warfare, and

(e) Space-to-Earth warfare.[43]

The "core of space warfare" is the struggle for information dominance, which is to be achieved by soft" and hard strikes on enemy space platforms, thereby, disrupting, destroying electronic equipment and computer system and fundamentally destroying his space-information system. According to the Chinese aerospace experts, space electronic warfare (EW) aimed at jamming, sabotaging, and destroying satellites is the most important way to gain information dominance in future wars. EW satellites travelling in geostationary orbits or 300-1,000 kilometre orbits can in fact conduct electronic reconnaissance and jamming in wide areas.[44]

Chinese military scientists assert that ASAT warfare is the most effective way to achieve space dominance. Aircraft, warplanes, and rockets can be used to launch anti-satellite missiles to destroy enemy satellites or "space landmines" can be placed on the orbits of enemy satellites for their destruction. In addition the strategists advocate use of lasers, clusters and microwaves weapons to attack enemy satellites. Based on the capabilities of reconnaissance satellites, Chinese aerospace scientists have compiled the following list of "space-information countermeasures":-

(a) Aim for the satellite's effective payload by applying suppression interference to cause overload in the satellite's receiving system, data processing system, and memory;

(b) Target the satellite's remote control system by;

    (i) Establishing a space target monitoring system to acquire the

---

[43] Ibid
[44] Ibid

satellite's technical parameters and character information, and

(ii) Effectively detecting and analyzing the satellite's operational system and down-link remote signal;

(c) Attack the satellite's space-to-ground communication and command nodes to weaken the connection, link, mutual operation, and networking flexibility in order to degrade its operational effectiveness; and

(d) Use high-energy and kinetic weapons to blind or destroy the Reconnaissance satellite while Chinese military experts applaud the "brilliant" performance of the U.S. Global Positioning System (GPS) in recent high-tech military operations, they continue to clarify its inevitable "Achilles' Heel." They have delineated three major weaknesses. First, defeat GPS at its source by exploiting the weakness of the low orbits of navigation satellites. This is accomplished by attacking with anti-satellite satellites, high energy laser weapons, and high-altitude weather-monitoring rockets. Second, defeat GPS by exploiting the scattered and exposed ground stations.[45]

Integrated Air-and-Space Operations. Chinese military strategists have long been articulating a body of operational concepts for conducting integrated "air-and-space operations" (ASO). The boundaries dividing military aviation and aerospace will gradually disappear to create a unified aviation and aerospace entity whose range extends from the surface of the Earth to the Outer Space. As experienced in the recent wars, the ground, air, and space already constitute an indivisible operational environment. Conducting of integrated ASO is no longer a matter of their feasibility but, now only a matter of perfecting the relevant technologies. Technological breakthroughs in systems such as the Space Shuttle, aerospace aircraft,

---

[45] China's Military Strategy in Space by Mary c. Fitzgerald. http://www.hudson.org/files/ publications/07_03_29_30_fitzgerald_statement.pdf

space weapons, and "new-concept" weapons, integrated ASO are becoming a new operational form of informationised warfare. For example, the Space Shuttle will become a completely new space weapon that combines aviation and spaceflight strikes, transportation, and information operations. This becomes Chinese future philosophy in their march for gaining the space dominance.

Motivation for the Chinese Space Quest

China's military modernization is driven by several factors that includes the desire to become a consequential global power; cope with regional security threats; respond to the American military presence around its periphery; and secure the critical access routes to its energy supplies.

Since China is confronted by formidable American military superiority, any effort to defeat the United States through an orthodox force-on-force encounter, centered on simple attrition, is doomed to fail. Chinese planners has concluded, that 'U.S. military forces, while dangerous at present, *are* vulnerable – and can be defeated by China with the right strategy'. Among many complex and diverse lessons, Chinese analysis of US military operations in the Persian Gulf wars, Kosovo and Afghanistan have yielded one critical insight: i.e., the United States is inordinately dependent on its complex but exposed network of sophisticated command, control, communications and computer-based intelligence, surveillance and reconnaissance systems operating synergistically in and through space. In other words, while American military power derives its disproportionate efficacy from its ability to leverage critical space assets, these very resources are simultaneously a font of deep and abiding vulnerability. Chinese strategists therefore, conclude that any effort to defeat the United States would require a riposte against its achilles heel: its space-based capabilities and their organic ground installations. An effective active defence against a formidable power in space requires China to have an asymmetric capability against the powerful United States. In essence, China will follow the same principles for space militarization and space weapons as it did with nuclear weapons. With this in view, China will develop anti-satellite and space weapons capable of effectively taking out an enemy's space system, in order to constitute a reliable and credible defence strategy.[46]

Space Policy of European Union

The European Union which consists of 27 member states has not yet formally issued a space policy. Currently, each member state pursues its own national space policy, though often coordinating through the independent European Space Agency (ESA). The EU is though a world leader in the technology does not play a prominent role in space as US and Russia. It uses space assets mainly for various civil applications. Jean-Jacques Dordain ESA's Director General (since 2003) outlined briefly the European Space Agency's mission: "Today space activities are pursued for the benefit of citizens, and citizens are asking for a better quality of life on earth. They want greater security and economic wealth, but they also want to pursue their dreams, to increase their knowledge, and they want younger people to be attracted to the pursuit of science and technology. I think that space can do all of this: it can produce a higher quality of life, better security, more economic wealth, and also fulfill our citizens' dreams and thirst for knowledge, and attract the young generation. This is the reason space exploration is an integral part of overall space activities. It has always been so, and it will be even more important in the future".[47]

European Union Initiatives

There is increasing debate within the EU about the role of space. High costs have discouraged development of independent networks by the EU members hence; the emphasis is more on linking national systems together. However, there are problems of interoperability for which many systems are being upgraded by the States of European Union.

---

[46] China's Military Space strategy by Ashley J Telis in www.carnegiceendows.org/files/tellis-china-space/.pdf

[47] European Space Agency, http://en.wikipedia.org/wiki/European_Space_Agency

However, with the United States for years enjoying a monopoly with its Global Positioning System (GPS), the European Union is increasingly getting concerned about America's dominance of such a critical system. In an effort to counter this US space dominance, the EU undertook to develop an independent capability and plans on launching the Galileo constellation of 30 satellites, and have it fully operational by 2013.[48] Galileo will be Europe's own global navigation satellite system, providing a highly accurate, guaranteed global positioning service which will avoid European Union members' dependence on US NAVSTAR GPS. However, it will be inter-operable with GPS and GLONASS, the two other global satellite navigation systems. The fully deployed Galileo system will consists of 30 satellites (27 operational + 3 active spares), positioned in three circular Medium Earth Orbit (MEO) planes at 23 222 km altitude above the Earth.

The other common initiative Global Monitoring for Environment and Security (GMES) mainly has civil applications and is aimed at streamlining European activities in the field of Earth observation. The system will aid the evaluation and implementation of European policies which have an impact on the environment.

Progress towards a European Space Policy and a common programme of activities is slow, partly because member states differ on issues such as, the importance of space relative to other priorities; whether Europe needs space assets independent of the US, and how the North Atlantic Treaty Organization would fit into any future EU programme; whether civilian systems may be used for other purposes. The UK is opposed to military use of these systems; while others including France, are not. There is also confusion over the roles of organizations involved in space activities. ESA is increasingly involved in security-related activities. Some favour this, while others say this conflicts with ESA's objective to promote peaceful uses of space, and point out that it has members who are not part of the EU. There is a dilemma of balancing the US and Europe. In fact, presently, the

---

[48] Ibid

EU Satellite Centre provides imagery and analysis to support EU decision-making on security and defence. It relies largely on commercial data but has some agreements with individual member states to use data from their military satellites.

France Space Policy

France contributes to the major infrastructure programmes of European Union, such as Galileo and GMES, as well as providing resources that help to give Europe independent access to space. It has helped Europe, to develop the Ariane launcher family. The new versions of Ariane V make it possible to launch large satellites. In addition to Ariane V, with the deployment of Vega and Soyuz at the Guyana Space Centre (GSC), Europe will be able to offer a complete line of launchers (light, medium and heavy), allowing it to meet operators' demands with the required flexibility. As such, France is participating in the ESA's compulsory programme, "Cosmic Vision", as well as the "Aurora" programme, for solar exploration and the robotic Mars exploration programme. It is also involved in implementing the European "Columbus" laboratory, docked with the International Space Station since February 2008.

The National Space Research Centre (CNES), which implements France's space policy, possesses all of the skills it needs to be able to manage all space civil and military applications. France has particularly advanced capability in the field of satellite imagery and has launched Helio 1 and Helio 2 dedicated military satellites and Syracuse 3 communication satellites.[49]

UK Space Policy

The British space policy's stated focus is to win sustainable economic growth, secure new scientific knowledge and provide benefits to all citizens.

---

[49] Defence Policy for the Defence of Europe,http://www.eurodefense.de/images/aspacepolicyfordefenceofEurope.pdf

At the same time, it aims to retain and grow a strategic capability in the space-based systems.[50] The space programme was earlier managed by the British National Space Centre which recently has been replaced by the newly formed UK Space Agency (officially launched on 23 March 2010). The UK Space Agency is responsible for all strategic decisions on the UK civil space programme. The agency will not directly manage security and military programmes but, will have a mechanism for interfacing with defence departments on these.[51]

The UK relies on the United States for most military space applications. There is no agency within the Ministry of Defence (MoD), and no specific budget, dedicated to military space. The Assistant Chief of the Air Staff within the Royal Air Force (RAF) co-ordinates space activities across the MoD. Activities are funded if they are a cost-effective way of achieving a specific objective. While the MoD's military space policy is classified, the RAF's Future Air and Space Operational Concept (FASOC) document indicates priorities over the next 20 years. FASOC highlights the role of small satellites and also the need for space surveillance (detecting and tracking objects in space). The MoD's space research priorities are small, low-cost satellites, surveillance of space systems, and space weather effects. The MoD spends £~1 million per year on space research. UK industry is developing expertise in small satellite technologies that could provide affordable means of gaining access to military space capabilities. Small satellites although they do not provide the same data quality as conventional satellites but, more can be deployed for the same cost, reducing reliance on a single satellite, allowing more timely data delivery and covering a wider area. Debate in the EU over the role of space in European security and defence policy is increasing. UK faces a challenge in striking a balance between collaboration with the US and EU.

---

[50] UK Space Agency. http://www.bis.gov.uk/ukspaceagency/what-we-do

[51] www.ukspaceagency.bis.gov.uk/About-Us/UK-Space-Agency/8002.aspx

Japan Space Policy

Japan issued its first basic Space Policy in May 2008. It undeniably did so to overcome crisis over Japan's use and R&D of space which suffered from serious shortcomings such as absence of general strategy in use of space and poor Japan's track record of space utilization. From 1969 to 1994 Japan did not have a successful satellite launch. Even 10 years later in 2004, Japans space programme has been described by some as undergoing the crises of confidence.[52] As on now Japan lacks the technological a scientific wherewithal to fabricate its own satellites and most of Japan's operational satellites such as broadcasting satellite are imported from overseas.

To overcome these weaknesses, Japan Legislature enacted a basic Space law on May 21, 2008 thus providing a major turning point for Japan's use and R&D of space. This law mainly applies to six fundamental principles; which are essentially are for peaceful use of space, viz; improvement of the lives of the people, development of industry, progress of human society, contribution to international activities and appropriate care of the environment. Promotion of advanced space development and utilization, promotion of education and learning etc. Law stipulates a review of policy in five years its formulation[53]

However, perceiving a security threat from North Korea Japan intends to use space for to meet the defence need particularly for reinforcement of information gathering and enhancement of warning and surveillance activities. The details of Japan's Defence Guidelines are not in public domain.

Japan also wants to work vigorously for promoting R&D of the future forefront areas and probe deep into space and expand the human's sphere of activities as a leading country of the space which may solve the

---

[52] Japans Space Programme : A Fork In the Road by Steven Berner, A Rand Study
[53] Japan's Basic Plan for Space Policy, http://www.kantei.go.jp/jp/singi/utyuu/basic_plan.pdf

worldwide environmental and energy issues confronting humankind, Japan has been involved in necessary research to source clean space solar power in about 10 years. Japan also wants to take lead to improve the space environment particularly to decrease the occurrence of debris caused by a launch of Japanese rockets and satellites and to increase the level of debris monitoring for preservation of the space environment in collaboration with the international society.

Japan, intends to carry out the research and development of ocean monitoring to respond to maritime events such as smuggling in Japan's surrounding ocean area, illegal operation of foreign fishing boats, suspicious vessels, major accidents at sea, or sea piracy on sea lanes. Needless to say these have also immense security applications. After the launch of North Korean missile Taep'o-dong in August 31, 1998, Information Gathering Satellites were introduced mainly for national security purposes.[54] In fact, in reaction to the perceived growing threat from North Korea, Japan announced, in August 2003, its intention to deploy, an advanced missile defence system, in contrast with the previous Japanese missile defence policy which used to stress only precautionary research and development in collaboration with the US.[55]

Space Policy of Israel

The Israeli Space Agency (ISA) was created in 1983 under the Ministry of Science and Technology. To date Israel's industrial base for launch vehicle and satellite development is narrow. Israeli industry is making substantial investments in space technology. Following the footsteps of India, Israel is first concentrating on the development of relatively simple launch vehicles with low payload capacity and of satellites based on proven technologies. Future activities may be biased towards the deployment of

---

[54] Ibid

[55] Background Paper: *Peaceful And Military Uses Of Outer Space: Law And Policy* Prepared by institute of Air and Space Law, faculty of law, Mcgill University, Montreal, Canada, Feb 2005.

more sophisticated space systems rather than a significant advance in booster capability.[56]

Survival of the state is an all pervasive national issue. Israel's short history is punctuated by sequence of serious wars. Its survival therefore depends on acquisition of "asymmetrical techniques" to offset the inherent serious disadvantages as also the numerical advantages of the adversaries. The principal nature of threat has shifted from the proximate land forces of Syria and Egypt to the long range missile capabilities of Syria and Iran. Under the prevailing circumstances the broad strategy of Israel is to construct comprehensive network of small spy Satellites in LEO to avoid being surprised at the same time maintain its own element of surprise capability. Israel strategy of small satellites using the micro and Nano-technology could fulfill its requirement of force enhancement. Israel strategic doctrine dictates that Space driven RMA is critical to future military Operation.[57] Appreciating an existential threat, Israel's initial investment in its space program is driven by strategic considerations, especially the ability to observe the activities of other states without violating international law. It is for this reason that the primary focus of Israel's space efforts has been and continues to be the development of high-resolution imaging capabilities and has made considerable advancement in developing Surveillance and observation satellites.

---

[56] Israel Space Agency, http://www.globalsecurity.org/space/world/israel/agency.htm

[57] *Space ,The frontier of modern defence*, 2006, by Squadron Leader KK Nair.

INDIA'S SPACE POLICY

Introduction

Space activities in the country started during early 1960s with the scientific investigation of upper atmosphere and ionosphere near Thiruvananthapuram using small sounding rockets. Realising the immense potential of space technology for national development, Dr. Vikram Sarabhai, the visionary leader envisioned that this powerful technology could play a meaningful role in national development and solving the problems of common man. Explaining the goals and objective of India's space activities, he stated, "There are some who question the relevance of space activities in a developing nation. To us, there is no ambiguity of purpose. We do not have the fantasy of competing with the economically advanced nations in the exploration of the moon or the planets or manned space-flight. But we are convinced that if we are to play a meaningful role nationally, and in the community of nations, we must be second to none in the application of advanced technologies to the real problems of man and society".[58] Explaining our quest for space technology, former President Dr.A.P.J Kalam stated, "Many individuals with myopic vision questioned the relevance of space activities in a newly independent nation, which was finding it difficult to feed its population. Their vision was clear if Indians were to play meaningful role in the community of nations, they must be second to none in the application of advanced technologies to their real-life problems." Thus the pioneer in space research in India had no intention of using the space a mean to display our might.[59]

[58] Indian Space Research Organisation, http://www.isro.org/

[59] http://en.wikipedia.org/wiki/Indian_Space_Research_Organisation

Overall Vision

Indian Space programme have mainly concentrated on achieving self reliance and developing capability to build and launch communication satellites for television broadcast, telecommunications and meteorological applications; and remote sensing satellites for management of natural resources. The indigenous development of satellites, launch vehicles and associated ground segment for providing these services are integral to those objectives. Accordingly, Indian Space Research Organisation (ISRO) has successfully operationalized two major satellite systems namely Indian National Satellites (INSAT) for communication services and Indian Remote Sensing (IRS) satellites for management of natural resources; also, Polar Satellite Launch Vehicle (PSLV) for launching IRS type of satellites and Geostationary Satellite Launch Vehicle (GSLV) for launching INSAT type of satellites.[60]

Space holds immense potential to accelerate the development process in the country and offers enormous opportunities to understand the universe; therefore, the thrust of our space programme is on large scale applications of space technology in the priority areas of national development. The future directions for Space Programme has taken into account needs of the country in the context of emerging international environment and the potential that India holds for human development. The broad direction for the space programme is:-

(a) To accelerate development of the country in key social and economic sectors.

(b) To achieve higher levels of self-reliance in critical areas of technology.

(c) Secure a unique place for the country by embarking on missions of strategic importance, and

---

[60] Indian Space Research Organisation, http://www.isro.org/

(d) In expanding knowledge about the universe, solar system and planet earth.

The policy does not state our ambition to use space to bolster the capabilities our armed forces. The defence technology in India is handled primarily by the Defence Research and Development Organization (DRDO). While ISRO which designs and launches Satellites, itself a civilian organization, doesn't have any interaction with defence forces or DRDO. In the past, however, there has been the criticism that expertise gained by ISRO has been used to develop Missile technology, by DRDO in its "The Integrated Guided Missile Programme (IGDMP)". Suspecting this, the U.S. State Department's in May 1993 imposed trade sanctions against ISRO for proliferating missile, specifically rocket engine technology. This action though delayed India's pursuit of space ambition but, in long run has been beneficial in achieving self reliance.Now, India has even indigenously developed the cryogenic launch technology which will be integrated as a the upper stage in the GSLV launch vehicle and help in launching the heavier satellites which hitherto were being launched with Russian or French launch vehicles.[61] (There are some glitches which eventually would get solved).

Organisation

Space Commission has the overall responsibility to implement the space policy of Government of India and formulates guidelines and policies to promote the development and application of space science and technology. In this activity the Space Commission is supported by other national level committees, such as INSAT Coordination Committee (ICC), the Planning Committee on Natural Resources Management System (PCNNRMS) and the Advisory Committee on Space Sciences (ADCOS).

The Department of Space (DOS), which was created in 1972 implements these programmes through mainly Indian Space Research

---

[61] Success of Cryogenic rocket launch will make India a leader in Rocketry, Economic Times New Delhi, Sunday, 11 April 10.

Organisation (ISRO), National Remote Sensing Agency (NRSA), Physical Research Laboratory (PRL), National Atmospheric Research Laboratory (NARL), North Eastern-Space Applications Centre (NE-SAC) and Semi-Conductor Laboratory (SCL). The Antrix Corporation, established in 1992 as a government owned company, markets the space products and services.[62]

Organisation Chart

[62] Indian Space Research Organisation, http://www.isro.org/

ISRO coordinates the space programmes related to different activities such as the development of satellite communication, earth observation, launch vehicles, space science, space-industry development and support to disaster management. ISRO is also active in international cooperation and other tasks related to the implementation and coordination of the national space programme.

Antrix Corporation Limited

The Antrix Corporation Limited is the commercial arm of DOS and is responsible for the marketing and international promotion and exploitation of products and services related to the Indian space programme. In particular, Antrix markets subsystems and components for satellites, undertakes contracts for satellites to user specifications, provides launch services and tracking facilities and other related services and activities. The commercial arm of ISRO, Antrix, is extremely successful in its role, which brings in substantial revenue to ISRO (for example it earned Rs 1,059 crore in 2008-09 and was accorded the status of mini-ratna by Govt of India in 2008).[63]

In close collaboration with ISRO, several specialised establishments operate under the responsibility of DOS. These establishments, located in various places all over the country, have responsibility in different fields of the space activity. The main space centres are:-

- Vikram Sarabhai Space Centre (VSSC) - specialised in the development of satellite launch vehicles and sounding rockets.

- ISRO Satellite Centre (ISAC) - the lead centre for satellite development, covering structures, thermal systems, spacecraft mechanisms, power systems and satellite integration.

- Satish Dhawan Space Centre (SDSC) - SHAR- Sriharikota Space Centre - India's prime launching pad facility, providing the

[63]   http://www.isro.org/pdf/Outcome%20Budget%202010-2011.pdf

launch infrastructure as well as solid propellant processing and
their testing. A second launch pad has been recently built at SDSC
SHAR.

- Liquid Propulsion Systems Centre (LPSC) - the lead centre in
  the area of liquid and cryogenic propulsion for launch vehicles
  and satellites.

- Space Applications Centre (SAC) - specialised in the development
  of payloads for communication, meteorological and remote sensing
  satellites; it conducts space applications research and development.

- ISRO Telemetry, Tracking and Command Network (ISTRAC) -
  it provides mission support to low-Earth orbit satellites and to
  launch vehicle missions.

- Master Control Facility (MCF) - the monitoring and control centre
  for the geo-stationary satellites.

- ISRO Inertial Systems Unit (IISU) - carries out research and
  development in inertial sensors and systems and allied satellite
  elements.

- National Remote Sensing Agency (NRSA) - an autonomous
  institution supported by DOS, it is responsible for acquisition,
  processing and distribution of data from remote sensing satellites,
  based in Hyderabad.

Overview of Space Programme[64]

India has developed several national space applications: in the area of
telecommunication and meteorological satellites the Indian National Satellite
(INSAT) is the Organisation responsible for the management and operation
of the fleet of in orbit satellites. INSAT is a joint venture between DOS,
the Department of Telecommunications (DOT), India Meteorological

---

[64] Indian Space Research Organisation, http://www.isro.org/

Department (IMD), All India Radio (AIR) and Doordarshan. INSAT currently operates satellites which contribute significantly to a variety of services in telecommunications and television broadcasting including meteorological observations, disaster communications, Tele-education, Tele-health.

In the area of Earth Observation, the Indian Remote Sensing (IRS) satellite system is the world's largest constellation of satellites in operation. It consists of nine satellites in total, IRS-1C and D, Resourcesat-1, IRS-P3, Oceansat-1, TES and the latest Cartosat-1, the advanced mapping applications satellite. Cartosat-2, Oceansat-2 and the Radar Imaging Satellite (RISAT).

Space applications satellites are used for development activities covering the entire Indian Territory. In the educational field, some pilot projects under the EDUSAT Programme have been started for schools, colleges and other levels of education. Primary school children are covered by the Educational TV (ETV) as well. GRAMSAT is an initiative to provide communications network for computer connectivity, data broadcasting, TV broadcasting and e-governance. Telemedicine is a further application of satellite communications in remote hospitals and health centres in distant villages.

In the field of search and rescue, India is a member of the international COSPAS-SARSAT programme for providing distress alert and position location service through LEOSAR (Low Earth Orbit Search and Rescue) satellite system.

In satellite Navigation, India decided to implement an indigenous Satellite-Based Regional GPS Augmentation System also known as Space-Based Augmentation System (SBAS) as part of the Satellite-Based Communications, Navigation and Surveillance (CNS)/Air Traffic Management (ATM) plan for civil aviation. The Indian SBAS system has been given an acronym - GPS and GEO Augmented Navigation

(GAGAN); the first GAGAN navigation payload has been fabricated and it is proposed to be launched in 2011. Two more GAGAN payloads will be subsequently flown, one each on two geostationary satellites, GSAT-8 and GSAT-10.2

The National Natural Resources Management System (NNRMS), under the aegis of DOS, carries out projects of data utilization of RESOURCESAT-1, launched in October 2003, for remote sensing and applications such as snow cover discrimination, multiple crop discrimination, crop condition assessment, surface wetness, delineation of soil mapping and salinity, discrimination of different forest types and crown densities, regional level land use/land cover information and updating of regional geologic mapping.

In space transportation activity, India made rapid progress in the design, development, manufacture and operation of two launch vehicle systems:-

- The Geosynchronous Satellite Launch Vehicle (GSLV), with a payload capability of 2000 kg, for launching communication satellites into GTO and

- The Polar Satellite Launch Vehicle (PSLV), with a payload capability of 1600 kg, for launching remote sensing satellites into polar orbits.

Three successful flights of GSLV have been made to date and progress is being made in the national development of Cryogenic Upper Stage (CUS) to replace the Russian-procured cryogenic stage on GSLV. The development of GSLV-Mk III is also on a positive path: it is a three-stage vehicle with a capability to launch 4 ton satellites into GTO.

In Space Science, India is preparing for two important scientific missions to take place Chadrayan-2 in 2013 and manned mission in 2016. Chandrayaan-1, India's first mission to Moon, was launched successfully on October 22, 2008 from Satish Dhawan Space centre (SDSC),

Sriharikota. The spacecraft was orbiting around the Moon at a height of 100 km from the lunar surface for chemical, mineralogical and photo-geologic mapping of the Moon. The spacecraft carried 11 scientific instruments built in India, USA, UK, Germany, Sweden and Bulgaria. After the successful completion of all the major mission objectives, the orbit has been raised to 200 km during May 2009. The mission was prematurely aborted but it had by then achieved 95% of its objective and made a most significant discovery of detecting trace of moisture on the surface of the moon. Chandrayaan-2 mission is planned to have an orbiter/lander/rover configuration. The mission is expected to be realized by 2013. The science goals of the mission is to further improve our understanding of origin and evolution of the Moon using instruments onboard Orbiter and in-situ analysis of lunar samples and studies of lunar properties (remote & direct analysis) using Robots/Rovers.[65]

There is also a Human Spaceflight Programme (HSP). The programme envisages development of a fully autonomous orbital vehicle carrying two or three crew members to about 300 km low earth orbit and their safe return. It is planned to realize the programme in 2015-16 time frame.

A significant result of the space programme is the development of national industrial capabilities in several areas of activity. The space industry partnership has allowed the participation of some 500 industries in small, medium and large-scale sectors either through procurement contracts, know-how transfers or provision of technical consultancy. Through its association with the space programme, Indian industry is able to develop advanced technology and to handle complex manufacturing of space systems and components. Considering the world potential for space markets and the role played by Antrix Corporation in the international marketing of space products and services, the activity of Indian space industries is expected to grow further.

---

[65] Indian Space Research Organisation, http://www.isro.org/

India's economic progress has made its space program more visible and active as the country aims for greater self-reliance in space technology. In 2009, India has launched 11 satellites, including nine from other countries—and it became the first nation to launch 10 satellites on one rocket. After development of the cryogenic engine India would be sixth nation to have this capability.

Conclusion

Former Indian President A.P.J. Kalam while addressing the Indian Air Force (IAF) president's fleet review held bold goals for India's aerospace future. He predicted that, by 2025, the IAF will be a model force for the rest of the world, able to "succeed in the electronically controlled warfare in the midst of space encounters, deep-sea encounters, and ballistic missile encounters." However, our space policy continues to emphasize the peaceful uses of space. Hence, one notices a kind of disconnect between the desires and actual practice.[66] In the eleventh plan period (2007- 12). The total outlay is Rs 39,750 crores. The objective of space activities continues to be non military aiming to improve the living conditions and general environment of the masses. The major goal of the Space Policy is enhance our capabilities in space communications and navigation, gain leadership in earth observation by improved imaging techniques through three thematic series of EO systems - Land & Water resources, Cartography and Ocean Atmosphere, development of advanced microwave imaging capability, strengthening ground systems and Undertake major applications projects in the area of Agriculture, land and water resource management, infrastructure and urban/rural development, etc.[67]

In Space and Science, the major thrust is realm of Space Transportation System, Operationalisation of GSLV Mk I11 with 4T launch capability. Perfect payload recovery and re-entry technologies. Conduct

---

[66] IAF will be Model for World by 20205: Kalam," *The Times of India*, March 7, 2007

[67] Report of the working group established by the steering committee of the planning commission11th plan .

demonstration flights of Reusable Launch Vehicle and gain expertise in critical technologies for Manned Mission. The advanced space science endeavours includes Chandrayaan, Multi-wavelength X-ray astronomy, Mission to Mars and Asteroid / Comet.

Annexure

Satellites Launched by India

Satellite	Launch Date	Launch Vehicle	Type of Satellite
RESOURCESAT-2	20.04.2011	PSLV-C16	Earth Observation Satellite
YOUTHSAT	20.04.2011	PSLV-C16	Experimental / Small Satellite
GSAT-5P	25.12.2010	GSLV-F06	Geo-Stationary Satellite
STUDSAT	12.07.2010	PSLV-C15	Experimental / Small Satellite
CARTOSAT-2B	12.07.2010	PSLV-C15	Earth Observation Satellite
GSAT-4	15.04.2010	GSLV-D3	Geo-Stationary Satellite
Oceansat-2	23.09.2009	PSLV-C14	Earth Observation Satellite
ANUSAT	20.04.2009	PSLV-C12	Experimental / Small Satellite
RISAT-2	20.04.2009	PSLV-C12	Earth Observation Satellite
Chandrayaan-1	22.10.2008	PSLV-C11	Space Mission
CARTOSAT - 2A	28.04.2008	PSLV-C9	Earth Observation Satellite
IMS-1	28.04.2008	PSLV-C9	Earth Observation Satellite

Satellite	Launch Date	Launch Vehicle	Type of Satellite
INSAT-4B	12.03.2007	Ariane-5ECA	Geo-Stationary Satellite
CARTOSAT - 2	10.01.2007	PSLV-C7	Earth Observation Satellite
SRE - 1	10.01.2007	PSLV-C7	Experimental / Small Satellite
INSAT-4CR	02.09.2007	GSLV-F04	Geo-Stationary Satellite
INSAT-4C	10.07.2006	GSLV-F02	Geo-Stationary Satellite
INSAT-4A	22.12.2005	Ariane-5GS	Geo-Stationary Satellite
HAMSAT	05.05.2005	PSLV-C6	Experimental / Small Satellite
CARTOSAT-1	05.05.2005	PSLV-C6	Earth Observation Satellite
EDUSAT (GSAT-3)	20.09.2004	GSLV-F01	Geo-Stationary Satellite
Resourcesat-1(IRS--P6)	17.10.2003	PSLV-C5	Earth Observation Satellite
INSAT-3A	10.04.2003	Ariane-5G	Geo-Stationary Satellite
INSAT-3E	28.09.2003	Ariane-5G	Geo-Stationary Satellite
GSAT-2	08.05.2003	GSLV-D2	Geo-Stationary Satellite

Satellite	Launch Date	Launch Vehicle	Type of Satellite
KALPANA-1(ME-TSAT)	12.09.2002	PSLV-C4	Geo-Stationary Satellite
INSAT-3C	24.01.2002	Ariane-42L H10-3	Geo-Stationary Satellite
Technology Experiment Satellite (TES)	22.10.2001	PSLV-C3	Earth Observation Satellite
GSAT-1	18.04.2001	GSLV-D1	Geo-Stationary Satellite
INSAT-3B	22.03.2000	Ariane-5G	Geo-Stationary Satellite
Oceansat(IRS-P4)	26.05.1999	PSLV-C2	Earth Observation Satellite
INSAT-2E	03.04.1999	Ariane-42P H10-3	Geo-Stationary Satellite
INSAT-2DT	January 1998	Ariane-44L H10	Geo-Stationary Satellite
IRS-1D	29.09.1997	PSLV-C1	Earth Observation Satellite
INSAT-2D	04.06.1997	Ariane-44L H10-3	Geo-Stationary Satellite
IRS-P3	21.03.1996	PSLV-D3	Earth Observation Satellite
IRS-1C	28.12.1995	Molniya	Earth Observation Satellite
INSAT-2C	07.12.1995	Ariane-44L H10-3	Geo-Stationary Satellite

Satellite	Launch Date	Launch Vehicle	Type of Satellite
IRS-P2	15.10.1994	PSLV-D2	Earth Observation Satellite
Stretched Rohini Satellite Series (SROSS-C2)	04.05.1994	ASLV	Space Mission
IRS-1E	20.09.1993	PSLV-D1	Earth Observation Satellite
INSAT-2B	23.07.1993	Ariane-44L H10+	Geo-Stationary Satellite
INSAT-2A	10.07.1992	Ariane-44L H10	Geo-Stationary Satellite
Stretched Rohini Satellite Series (SROSS-C)	20.05.1992	ASLV	Space Mission
IRS-1B	29.08.1991	Vostok	Earth Observation Satellite
INSAT-1D	12.06.1990	Delta 4925	Geo-Stationary Satellite
INSAT-1C	21.07.1988	Ariane-3	Geo-Stationary Satellite
Stretched Rohini Satellite Series (SROSS-2)	13.07.1988	ASLV	Earth Observation Satellite
IRS-1A	17.03.1988	Vostok	Earth Observation Satellite
Stretched Rohini Satellite Series (SROSS-1)	24.03.1987	ASLV	Space Mission

Satellite	Launch Date	Launch Vehicle	Type of Satellite
INSAT-1B	30.08.1983	Shuttle [PAM-D]	Geo-Stationary Satellite
Rohini (RS-D2)	17.04.1983	SLV-3	Earth Observation Satellite
INSAT-1A	10.04.1982	Delta 3910 PAM-D	Geo-Stationary Satellite
Bhaskara-II	20.11.1981	C-1 Intercosmos	Earth Observation Satellite
Ariane Passenger Payload Experiment (APPLE)	19.06.1981	Ariane-1(V-3)	Geo-Stationary Satellite
Rohini (RS-D1)	31.05.1981	SLV-3	Earth Observation Satellite
Rohini (RS-1)	18.07.1980	SLV-3	Experimental / Small Satellite
Rohini Technology Payload (RTP)	10.08.1979	SLV-3	Experimental / Small Satellite
Bhaskara-I	07.06.1979	C-1 Intercosmos	Earth Observation Satellite
Aryabhata	19.04.1975	C-1 Intercosmos	Experimental / Small Satellite

Source: ISRO website

SPACE AS A STRATEGIC ASSET

Introduction

Today, besides, enormous amounts of space debris, there are some 800 active satellites in orbit around the world. Of these 800, 68 percent are for communications – some military and other available to the military during war, 7 percent are for navigation and surveying, 6 percent are spy satellites observing in all frequencies, 5 percent are weather and oceanographic satellites and other 5 percent to do the miscellaneous works like tracking elephants and polar bears etc. If satellites are meant to gain the greatest use of space for the human kind, then it is relevant to note that 760 satellites out of 800 look at the earth, while remaining 40 look outwards at the outer space. So, the most useful segment of space that man is concerned is not inter–stellar discovery or space travel but, for gaining an advantage, competitively on the earth.[68]

The great amount of money, effort and organisation in space research has gone into making possible communication satellites, spy satellites, navigation and weather satellites and rocketry. These activities are for terrestrial supremacy and not for space exploration. There was a time when nuclear weapons and their effective delivery mean marked out nations as great power. Today there is little doubt that state power projection and the Revolution in Military Affairs (RMA) will be based on enormous wide band connectivity run through satellite communication systems. The information that is passed through these channels will contain geographical

---

[68] An article titled "Strategic space" by Raja Menon, in Space Security and Global Cooperation, 2009

positions particularly about enemy dispositions, obtained from surveillance satellites. The entire architecture for imposing force upon a state has become heavily dependent on the use of satellites. The US is currently supreme in this area, but it is an area in which China will increasingly play a challenging role. Hence, outline of another competitive race is already discernible.[69]

The perception of strategic value of space took place after the collapse of the Soviet Union, when the United States appeared as a leading space power, a long way ahead of other states. The growing role of space systems during the 1990s was largely reinforced by the RMA. In this, there is an innovative application of space technologies resulting in fundamental changes in the nature of warfare. United States use of large constellation capabilities in recent conflicts has offered a perfect illustration of this new approach.[70]

At a strategic level, space assets are used for arms control verification and early warning systems. But, space also offers capability for linking vast distances, gathering information, improving and expanding the coverage of education, expanding medical services, monitoring and managing environmental issues etc. These capabilities clearly are of value to all nations. Besides, a space asset apart from so called strategic level is more often employed in the military at the tactical level.[71]

Importance of Space Assets

Space assets virtually affect every aspect of our life. These have dramatically changed and improved our lives. We may reflect on how our life would be without them. What if INSATs malfunction or simply are not available? The financial sector would get affected as automated teller machines wouldn't process transactions; television and radio transmissions would get interrupted, our communication facilities would come to naught.

[69] ibid

[70] *Strategic space a variable geometry concept* , an article by Isabelle Sourbes-verger, Space Security and Global Cooperation, 2009

[71] Space as strategic Asset by Joan Johnson Freese, Page 6, 2007, Columbia University Press

Inter and intra state communication would be badly affected or simply may not exist. The governance and administrative machinery would simply not function at its optimum.

A State can also not conduct modern diplomacy without space assets as we communicate overseas in real time via satellite communications. What if more satellites are destroyed? The move of the nation practically would come to grinding halt. It is thus not difficult to estimate the impact of space-based assets on our daily lives.With development in space technology, the reliance on these space-based assets is only increasing with time.

The dependence of developed countries on their space assets is much more than developing or underdeveloped country. United States for example, uses space assets to support war fighting. Its Defence Department has long acknowledged the vital importance of various critical national security space assets which the United States employs such as the Defence Support Programme and Space Based Infrared System which perform early warning surveillance, detection and ballistic missile tracking functions; and the Defence Satellite Communications System that provides secure voice and data communications to forces which are deployed all around the world. The reliance on these space-based assets is only increasing.

While space assets are vital to national security and foreign policy, they are equally and increasingly important to the global economy and welfare. Banks use the highly accurate timing signal from the on-board atomic clocks to guarantee all offices record transactions simultaneously around the world. Space systems, services, and capabilities provide weather forecasting, enable search and rescue missions, and provide for emergency communications. Space-based monitoring of the Earth's crust to better aid in earthquake prediction, environmental protection is improved by the characterization of contaminated soil, sediment, and water sites, and space-based satellites ensure that international air and sea traffic arrive safely and navigate accurately.

Vulnerability of Space Assets

Space laws do not prohibit militarisation of the space, nor do they prevent space vehicles from carrying conventional ordnance. It is also accepted that satellites today are the basis for the information passing architecture of economies and militaries. Hence, the dependence on satellites constitutes terrible vulnerability that is bound to be exploited by a country which feels geopolitically threatened. Space systems are, by their very nature, vulnerable to a range of threats. The prevailing security environment is giving clear signal that in time to come; China will increasingly play a challenging role. In fact, On 11 January 2007, a Chinese medium-range ballistic Missile carried out a kinetic kill of an ageing Chinese weather satellite deployed in low Earth orbit at an altitude of some 864 kilometres. The satellite was heading south at the time of its intercept, and was intercepted approximately on a virtual head-on collision mode at extremely high velocity. The test highlighted Chinese space capabilities as well as the vulnerability of space assets.

The threats also include jamming satellite links or blinding satellite sensors, which can be disruptive or can temporarily deny access to space-derived products. Anti-satellite weapons whether kinetic or conventional or Electro-Magnetic Pulse (EMP) weapons can permanently and irreversibly destroy satellites. Military force can be employed against ground relay stations, communication nodes, or satellite command and control systems to render space assets useless over an extended period of time. Adversaries can also employ denial and deception techniques to confuse or complicate our information collection.

Moreover, the ability to restrict or deny our freedom of access to, and operations in space is no longer limited simply to nation states. With knowledge of space systems, their orbits, and the means to counter them being readily available, both state and non-state actors can acquire or develop knowledge about our systems, their capabilities, and how to disrupt or destroy them. For instance, non-government satellite observers track satellites and post their orbits on the internet. Terrorist groups might employ GPS jammers; or our ground stations and communications nodes could be disabled or destroyed by terrorists using, for example, rocket-propelled grenades. Terrorists, like state adversaries, understand our vulnerabilities and have targeted our economy in the past. Thus, our space infrastructure could be seen as a highly lucrative target and today more actors have greater access to increasingly sophisticated technologies and capabilities that will improve their ability to interfere with space systems, services, and capabilities.

The Outer Space Treaty which was drafted more than 40 years ago, is as relevant and applicable today as it was then. It has established the guiding principles for space operations by which we believe all nations should conduct themselves. A quick look at some of the treaty's key provisions shows: that space shall be free for all to explore and use; that space activities shall be carried out in accordance with international law, including the Charter of the United Nations, which guarantees the right of self defence. The treaty also prohibits placing weapons of mass destruction in orbit and prohibits the parties from interfering with the assets of others.

Not all countries can be relied upon to pursue exclusively peaceful goals in space. A number of countries are exploring and acquiring capabilities to counter, attack, and defeat space systems. United States and erstwhile Soviet Union and china have already displayed their Space offensive capability. Unimpeded access to and use of space is of vital national interest to us. In view of the growing threats, our space policy requires review to increase our ability to protect our critical space capabilities and to continue to protect our interests from being harmed through the hostile use of space. Ignoring this vital area could prove costly. We have already paid for our complacency in 1962 India china war. Space will always remain an attractive option for the adversary because the political, military, and economic value of space assets. Ensuring the freedom of space and protecting our interests in this medium should therefore become national security priority.

Both Russia and china continue to make persistent demand for negotiating treaty for Prevention of Arms Race in the Outer Space commonly known as PAROS (most recently at Shanghai Cooperation Organization meet in 2009) which United States is unwilling, terming it as a hypothetical future arms race in space as weapons are still not deployed in the space. It does not favour negotiating for more international constraints which are not verifiable. Moreover, the proposal of PAROS appears to be more of a tactical ploy by both Russia and China to seek more time to seek space technology parity with United States. If this was not true what motivated them to develop anti satellite weapons while at the same time seeking PAROS treaty.

To ensure free access to space, however, we must continue to develop a full range of options to deter and defend against threats to our space infrastructure. Deterrence requires first and foremost a clear statement of what interests are vital. We must be clear that protecting space assets is of vital national interest. No nation, no non-state actor, should remain under a illusion that India will tolerate a denial of our right to the use of space for peaceful purposes and oppose others who wish to use their military capabilities to impede or deny our access to and use of space. We should seek the best capabilities to protect our space assets by active or passive means particularly in view of inimical neighbours that surround us. These means include non-space back-ups, on-board sub-component redundancy, manoeuvring, system hardening, encryption, and frequency agility etc.

Conclusion

Our National Space Policy is committed to peaceful uses of space for the benefit of all citizens. It must also take in to account the emerging realities in Space. If United States, Russia and China have taken steps to safe guard their space assets why should we not develop means to maintain freedom to use our Space asset and be vigorous in defending them?

MILITARISATION OF SPACE

Acquisition of high grounds for military advantage has been a perennial feature of military campaigns. For thousands of years, military tacticians have exploited the concept of "capturing" or "keeping" the high ground in military campaigns. The first form of man-made flying objects were kites. The earliest known record of kite flying is from around 200 B.C. in China, when a general flew a kite over enemy territory to calculate the length of tunnel required to enter the region. During the First World War there are instances of using the higher platforms to gain intelligence of the enemy. For example hot air balloons were also lofted by Napoleon to observe troop movements.

Planes almost as soon as they were invented were used by the military. The first country to use planes for military purposes was Italy, whose planes made reconnaissance, bombing and shelling during the Italian-Turkish war (September 1911 – October 1912), in Libya. In the First World War, the Allies and Central Powers both used planes extensively in offensive, defensive and reconnaissance capabilities. Aircraft revolutionized warfare during the twentieth century, leading to "command of the air" as a key strategic concept. Following the shooting down of high altitude aircraft like the U-2, the quest for safer observation went further into space. Initial attempts for control of the environment of space were led by both the US and the Soviet Union.[72]

Presently, the use of outer space is developing in two directions. On

---

[72] http://en.wikipedia.org/wiki/Militarisation_of_space

the one hand, space technology is used in every corner of contemporary human life; on the other hand, the world has experienced accelerating steps of outer space militarisation. The traditional military use of outer space has spread from supportive roles such as communication, navigation, reconnaissance, surveillance and early warning at peacetime, to direct war fighting roles such as command and control, target positioning and bomb guiding. Today, judging by the direction and pace of development, the weaponisation of Outer space appears imminent.

Military Satellites

A military satellite is an artificial satellite used for a military purpose, often for gathering intelligence, as a communications satellite used for military purposes, or as a military weapon.[73] The first so called "military satellite" KH-11 was launched by United States in the space in 1976. It was equipped with large telescopes and video cameras to observe Earth and continuously transmit pictures to the ground stations. As was natural in the cold war, Soviet Union in response to the United States KH-11, developed and launched Cosmos in 1982. Its image quality was poor than KH-11 but, was produced and launched in more numbers. Later by 1988, Israel and South Africa also constructed and launched their Military Satellite Offea 1, followed by France, Italy and Spain.

A satellite by itself is neither military nor civil. Payload of the satellite decides its military or civilian character. Even this distinction is now blurred. For example, a civilian satellite can carry military transponders and vice versa. Civil commercial satellites are also known to carry out military tasks such as providing military communications, imagery etc. For example, during military operations in Kosovo in 1999 and Iraq in 2003, eighty percent of all space borne communications were transmitted through multinational and commercially owned satellites, and such reliance on commercial satellites continues to grow. The increased reliance by the

---

[73] Ibid

US on commercial satellites also prompted General Richard Meyers, the Chairman of the Joint Chiefs of Staff to state: "Clearly, our reliance on commercial space has created a new centre of gravity that can easily be exploited by our adversaries".[74] At the same time, military satellites like the NAVSTAR GPS have more civilian users than military users. The US and Russia lead on military space activities, but more countries are now getting involved. The dual use commercial satellites are possibly the reason for the growth. US, Russia, China, India, Israel, Japan and European Space Agency (ESA) have launch facilities, which other countries can pay to use.

Uses of Military Satellites

The military uses of satellites include:-

(a) **Imagery.** To perform tasks such as identification of targets, battle damage assessment, detect effect of nuclear detonations etc.

(b) **Navigation and Targeting.** Navigation satellites and GPS are used to guide military actions during operations. The best example of military satellite is the NAVSTAR GPS (Navigation Signal Timing and Ranging Global Positioning System). The NAVSTAR GPS network is operated by US Air Force. It contains an accurate and reliable satellite navigation system that determines the position of the military forces on the ground, air or at sea and provides coordinates for accurate targeting of the enemy positions. The network is also used for many commercial purposes. Its 25 spacecraft which provide the wide spread coverage including the North and South poles. The space craft are in semi synchronous orbits inclined at 55 degrees to the equator at 18700 km (11600 miles) altitude. Russia, European Union and China are also launching their versions of GPS. India would also have its own indigenous Indian Regional Navigation System (IRNS) comprising seven

[74] Foes See U.S. Satellite Dependence as Vulnerable Asymmetric Target, *JINSA Online*, December 04, 2003 http://www.globalsecurity.org/org/news/2003/031204-jinsa.htm

satellites operational by 2014.

(c) **Signal Intelligence (SIGINT).** SIGINT is used to detect radio transmissions and to listen enemy communications. Early warning Satellites are used to sense the launch of the ballistic missiles and nuclear detonations.

(d) **Telecommunications.** In military operations, this enables exchange of information, for example between the "frontline" and strategic commanders, so that decisions are based on up to date intelligence.

(e) **Early Warning.** Satellite sensors can spot missile launches by detecting their hot plumes. However, the technology to *track* missiles along their trajectory, from space, is in its early stages of development.

(f) **Meteorology.** To provide weather data for the military. The Weather data is needed to plan and successfully execute operations.

(g) **Geodesy.** Military satellites are also used for Geodesy, i.e. the study of Earth's shape and size. Data from geodesic surveys is important to the military, as it is used for map making, positioning, navigation, and for a variety of other missions.

Militarisation vs. Weaponisation of Space

Since the beginning of the space era, the world community has strongly endorsed the use of outer space for "peaceful" purposes. Although the term appears in many UN documents and space law treaties, the term "peaceful" still lacks universally accepted definition. Initially, common interpretation of the term "peaceful" in relation to outer space was "non-military." However, the term "peaceful uses" is now considered to mean "non-aggressive" rather than "non-military." Now most space faring nations sharing the original interpretation of "peaceful" appears to have accepted that outer space may also be used for military purposes.[75]

In this context, a distinction must be made between "militarisation" and "weaponisation" of outer space. If one accepts the position that militarisation of space began with the launching of the earliest communications satellites serving military objectives, weaponisation is generally understood to refer to the placement in orbit of weapon systems that could attack targets in space or on the Earth. Although to this day there is no authoritative definition of "space weapon", there are space-based devices that have indirectly a destructive capacity (e.g., satellites serving GPS navigation of military aircraft and precision guided missiles). Till date, satellites themselves have no destructive capacity on their own and their support of military missions is not considered weaponisation of space[76].

According to a definition proposed by a group of United Nation Institute of Disarmament and Research (UNIDIR) experts: "A space weapon is device stationed in outer space (including the Moon and other celestial bodies) or in the earth environment designed to destroy, damage or otherwise interfere with the normal functioning of an object or being in outer space, or a device stationed in outer space designed to destroy, damage or otherwise interfere with the normal functioning of an object or being in the earth environment. Any other device with the inherent capability to be used as defined above will be considered as a space weapons."[77] While this type of definition includes also ground, sea and air based weapons in the category of space weapons, more recent definitions refer mainly to space based weapons.[78]

---

[75] Background Paper"*Peaceful And Military Uses Of Outer Space:Law and Policy*" Prepared by Institute of Air And Space Law,Faculty of Law, Mcgill University, Montreal, Canada February 2005.

[76] ibid

[77] http://praxis.leedsmet.ac.uk/praxis/documents/space_weapons.pdf

[78] /www.spacedebate.org/definition/Space%20Weapon/

Justification for Weaponisation

During the Cold War the United States and Soviet Union both relied upon satellites for command and control of nuclear forces. Destruction of these satellites could render one's nuclear forces vulnerable to a first strike. Spy satellites were used by militaries to take accurate pictures of their rivals' military installations. Despite the conclusion of the Cold War, the Cold War strategic anxieties continue to dictate US, Russian and Chinese nuclear forces. It is feared that destruction of satellites would severely affect military operations, possibly increase casualties. These strategic concerns have fuelled desire of leading space faring nations to weaponise space. Post Cold War space militarisation seems to revolve around three types of applications:

(a) The first application is the continuing development of "spy" or reconnaissance satellites which began in the Cold War era, but has progressed significantly since that time. Spy satellites perform a variety of missions such as high resolution photography (IMINT), communications eavesdropping (SIGINT), and covert communications (HUMINT). These tasks are performed on a regular basis both during peacetime and war operations. Satellites are also used by the nuclear states to provide early warning of missile launches, locate nuclear detonations, and detect preparations for otherwise clandestine or surprise nuclear tests (at least those tests or preparations carried out above- Early-warning satellites can also be used to detect tactical missile launches; this capability was used during Desert Storm when America was able to provide advanced warning to Israel of Iraqi SS-1 SCUD missile launches.

(b) The second application of space militarisation currently in use is Global Positioning System (GPS). The US military refers to it as NAVSTAR GPS - Navigation Signal Timing and Ranging Global Positioning System. This satellite navigation system is used for determining one's precise location and providing a highly accurate time reference almost anywhere on Earth. It uses an Intermediate Circular Orbit (ICO) satellite constellation of at least 24 satellites. The GPS system is designed and controlled by the United States Department of Defence and can be used by anyone, free of charge. The cost of maintaining the system is approximately US$400 million per year, including the replacement of aging satellites. The first of 24 satellites that form the current GPS constellation was placed into orbit on February 14, 1989. The primary military purpose is to allow improved command and control of forces through improved location awareness, and to facilitate accurate targeting of smart bombs, cruise missiles, or other munitions. The satellites also carry nuclear detonation detectors, which form a major portion of the United States Nuclear Detonation Detection System. On May 1, 2000, US President Bill Clinton announced that "Selective Availability" in a war zone or global alert while allowing military units to use GPS. However, European concern about the level of control over the GPS network and commercial issues has resulted in their planning the Galileo positioning system. Russia is also operationalisng an independent positioning system called GLONASS (global navigation system),

(c) The third current application of militarisation of space can be demonstrated by the emerging military doctrine of network-centric warfare. Network-centric warfare relies heavily on the use of high speed communications which allows all soldiers and branches of the military to view the battlefield in real-time. Real-time technology improves the situational awareness of all of the military's assets and commanders in a given theatre. For example, a soldier in the battle zone can access satellite imagery of enemy positions two blocks away, and if necessary mail the coordinates to a bomber or weapon platform hovering overhead while the commander, hundreds of miles away, watches as the events unfold on a monitor. This high-speed communication is facilitated by a separate internet created by the military for the military. Communication satellites hold this system together by creating an informational grid over the given theatre of operations. The Department of Defence is currently working to establish a Global Information Grid to connect all military units and branches into a computerized network in order to share information and create a more efficient military.[79]

Is Weaponisation of Space Real?

The Outer Space Treaty of 1966 prohibited the signatories' of treaty in placing of nuclear weapons or any other weapons of mass destruction in orbit of Earth, installing them on the moon or any other celestial bodies , or to otherwise station them in outer space. It however, does not forbid placing of conventional weapons in the outer space. Currently, military operations in space primarily concern either the vast Tactical advantages of satellite based surveillance, communications, and positioning systems or mechanisms used to deprive an opponent of said tactical advantages. A communications or reconnaissance satellite though provide a force multiplier effect in warfare with enhanced tactical situational awareness and resultant better command and control but, cannot be classified as a weapons as these are in war supporting role and don't technically violate the international law on the space which forbids placing of nuclear and conventional weapons of mass destruction. Similarly, transit of ICBM through the outer space on track to their targets in other continents or attack on satellites may come in the ambit of space warfare but do not constitute the weaponisation of space. Shooting down of P78-1, a communication satellite in a 345 mile (555 km) orbit by US F 15 in mid 1980, destruction of one of its obsolete satellite by China in 2007, and United States in 2008 are thus the only instances of space warfare but do not constitute weaponisation of the space as the F15 ac and the missiles

---

[79] http://en.wikipedia.org/wiki/Militarisation_of_space

were launched from the Earth.

The Future

Former United States Vice President Lyndon Johnson predicted in the beginning of the space race that," Control of space means control of the world." Same sentiments have been expressed by Commander-in-Chief of US Strategic Command, "It's politically sensitive, but it's going to happen. Some people don't want to hear this, and it sure isn't in vogue, but — absolutely — we're going to fight in space. We're going to fight from space and we're going to fight into space. That's why the US has development programs in directed energy and hit-to-kill mechanisms. We will engage terrestrial targets someday — ships, airplanes, land targets — from space." Similarly, General Xu Qiliang, head of PLAAF, in 2009 prophesied that, "The competition between military forces is moving towards outer space ... this is a historical inevitability and a development that cannot be turned back".[80] The proponents of space weaponisation argue that developing and maintaining core space control functions of situational awareness, defensive counterpace and offensive counterpace would assure freedom of action in space and which would enhance war fighting capability. Space superiority is becoming now what in the past was air superiority for a successful military mission. It is argued that if nation has legitimate interests in space, it should also be free to protect its access to space and its both civilian and military assets in space. The apparent vulnerability of the space assets, make them a very appealing target for a Space. The logic is that the greater economic and military dependence on space assets, the greater the adversaries will have motivation to destroy or at least cripple these economic and military advantages.[81] Therefore, any nation with space assets, when faced with the threat of survivability of its satellites might consider deploying weapons

---

[80] http://en.wikipedia.org/wiki/Militarisation_of_space

[81] The military use of space , a diagnostic assessment by Barry D Watts Centre for Strategic and Budgetary Assessment , Feb 2001.

in orbital space as an effective counter measure or deterrent.[82]

The opponents of weapons in space argue that the flight-testing, deployment and use of space weapons would pose a significant threat not only to military uses of outer space, but also to space exploration and other peaceful uses of space. Furthermore, using anti-satellite weapons would aggravate the problem of space debris, According to a report by UNESCO made public in London on 28 April 2002; there were about 2.7 tons of various missile fragments in orbit. US Space Command's Space Catalogue currently tracks some 9,000 man-made objects in orbit, ranging in size from 10 cm in lower orbit to over 1 meter in geo-stationary orbit. Approximately 94 per cent of these objects are considered space debris and a hazard to satellites and other spacecraft. With this material flying at speeds of almost 8 km/s (which is ten times more than that of a rifle bullet), a collision with a functional space object would cause it serious damage or even destruction. Weaponisation of space would only worsen the debris problem and could jeopardize the possibility of further space explorations and severely impair both civilian and military uses; it is also argued that space weapons could become "sitting ducks in orbit", thus create a new weakness, not a new strength. Satellites as such are already becoming a weak "center of gravity" in US military planning since they are vulnerable to electronic jamming, orbiting debris and electromagnetic pulse. The space based weapons would be costly in fact, ground based options can be just as effective against the space based threat. In short, space weaponisation has the potential to disrupt international stability and may negatively impact military strategic balance, alliance ties and relations among major powers. In such a case it may actually create more threats than provide protection for space assets.[83]

---

[82] http://www.csbaonline.org/4Publications/PubLibrary/R.20010201.The_Military_Use_o/ R.20010201.The_Military_Use_o.pdf

[83] Background Paper"*Peaceful And Military Uses Of Outer Space:Law And Policy* " prepared by Institute Of Air And Space Law,Faculty Of Law, Mcgill University, Montreal, Canada February 2005.

INDIA'S MILITARY CAPABILITY IN SPACE

The modern war fighting capabilities are getting highly dependent on the space assets and technologies whether it be real time situational awareness or for precision guidance of stand off weapons. The trend toward net centric warfare is unstoppable since it offers hitherto unavailable military advantage to technically superior forces and satellite based sensors and communication link which constitute such net work capabilities. Indian Defence forces as on date, however, has limited access to the space technologies. Unlike most other space faring nations Indian space programme has evolved entirely for capabilities in the civilian space domain, creating independent assets for peaceful applications. The changing global space will require significant R&D in critical technologies that could contribute to the military space capabilities. Hence, with in the country, there is an urgent need for an informed debate on the subject to prioritize action for space security of India.

The Defense Research and Development Organization (DRDO) is the primary body in India that develops space technologies for military applications. The current cooperation between the DRDO and ISRO is limited because ISRO mandate is to develop civilian space programs. Nonetheless, ISRO has been under criticism that there has been cooperation between the two institutions in the past. In support of this belief it is often stated that India's ballistic missile program is have grown out of experience gained in space launch vehicle (SLV) projects. Pursuing this belief, the U.S. State Department imposed trade sanctions in May 1992 against ISRO for proliferating missile, specifically rocket engine technology.[84] US has recently removed ISRO and DRDO and their subsidiaries from the entity list after President Obama India's visit in Dec 2010. This move may give India access to high end space technology which hitherto was not available.

The Indian military has no dedicated satellites for exclusively military operations.[85] However, certain ISRO satellites are dual-use and therefore can be used for both civilian and military applications. ISRO has focused its efforts on two major activities, ie, satellites used for remote sensing, meteorology and communications and rockets and launch capabilities for its satellites. These admittedly are meant for the civilian use but, are also useful for the military. For instance, the one-meter, high-resolution Technology Experiment Satellite (TES) launched in 2001, and the 2.5 meter, high-resolution Cartosat-1 launched in 2005 could provide imagery of the given area when required by the military. Similarly, the Cartosat-2A satellite, which was launched in April 2008 includes Israeli synthetic aperture radar technology has one meter resolution and was reportedly meant for dedicated military use.[86] This possibility however, has been refuted and it is stated that satellite will be used for civilian urban mapping purposes. Another satellite RiSat-2 launched in Apr 2009 from Sriharikota as been reported in the media as a spy satellite primarily to keep an eye on its borders round-the-clock and help the government in anti-infiltration and anti-terrorist operations. The 300-kg radar-imaging satellite has been built by Israel was launched India's home-grown rocket, the Polar Satellite Launch Vehicle (PSLV). This remote-sensing advanced imaging satellite has been positioned 550 km above the earth and it has all-weather capabilities. It carries Synthetic Aperture Radar (SAR) payload, which can take images during day, night and all weather conditions including under cloud cover, a capability that Indian satellites do not have. The

---

[84] http://www.defenseindustrydaily.com/india-building-a-military-satellite-reconnaissance-system-0996/

[85] http://www.isro.org/satellites/allsatellites.aspx, all satellites

[86] http://en.wikipedia.org/wiki/Cartosat-2A

significance of the satellite is its all-weather capability. It will be primarily used for defence and surveillance. The satellite also has good application in the area of disaster management and in managing cyclones, floods and agriculture-related activities.[87] Officially, the Cartosat platforms will be used for cartographic purposes, as well as urban and rural development. Unofficially, they are effectively dual-use even though they fall short of the 10-15cm (4"-6") capabilities of the best military satellites today.[88]

Priority Areas

The use of satellites to expand and enable military capabilities has attracted attention of the strategist in India for some time; two major military priorities emerge out of India's current space capabilities. One is extension and improvement of satellite reconnaissance capabilities for the defence and civilian use, and the second is an integrated Aerospace Defense Command.[89] An integrated Space cell was established in the Integrated Defence Head Quarters in Jun 2008 as forerunner to the Tri-service Space Command by Ministry of Defence as nodal agency to oversee the space security needs. The decision was announced by Defence Minster AK Anthony in combined Commander Conference to counter the "offensive counter space systems like anti-satellite weaponry, new classes of heavy-lift and small boosters and an improved array of Military Space Systems have emerged in our neighbourhood," after China's ASAT test in Feb 2007.[90] Both pursuits have been on the Indian agenda since the late 1990s. The primary motivations originate in the 1999 Kargil conflict and continued instability with Pakistan, coupled with India's rising regional role. If India had "harnessed space capabilities for national security" in 1999, enemy incursions could have been detected and assessed with more

---

[87] http://www.india-defence.com/reports-4320

[88] http://www.defenseindustrydaily.com/india-building-a-military-satellite-reconnaissance-system-0996/

[89] *Militarisation of Space : Security implications* By Amitav Malik, *CLAWS journal, winter 2008.*

[90] *Now a space cell to keep an eye on China's plan,*Times of India ,11 Jun 2008

accuracy; air power could have been applied with greater precision; and the loss of life could have been minimized.[91] It is likely that dual-use programmes in terms of ASAT capabilities and space weapons will be pursued in order to keep pace with potential adversaries. India's status as a rising Asian power point to the history of the nuclear Non-Proliferation Treaty, in which India lost out as a "legitimate nuclear weapon" status, hence, it is opined that the time to declare possessing an ASAT capability is now, before any treaties or regimes present restraints. This is significant considering that V.K. Saraswat, the DRDO Chief and Scientific Advisor to the defence minister asserted on Feb 10 after successful Agni III tests that, "We have the capability for interception of satellite. But we do not have to test because it is not our primary objective. There are repercussions of satellite interception like debris flying in the space".[92] This is not surprising as India is already in the threshold of operationalisng indigenous ballistic missile defence programme which could become the base for future development of ASAT weapon as and when decision is taken to develop this system.

Joint Co-operation in High End Technologies

India is also persuing agreements with other countries to expand its access to sophisticated satellite imagery for military uses. In this regard, there is a significant cooperation in sharing the satellite technology with Russia and Israel. France too has significant capability in remote sensing and imagery and is of friendly disposition to India and can become a dependable partner in developing high end technologies.

Israel. India and Israel have co-operated on a number of ventures. India's Satellite launch capabilities complement Israel's developments in remote sensing satellite technology. Israel's synthetic aperture radar allows a satellite to "see through" clouds, rain and other adverse conditions and

---

[91] Space the frontier of modern defence By KK Nair

[92] http://www.thaindian.com/newsportal/uncategorized/india-has-anti-satellite-capability-drdo_100318100.html

high resolution pictures using radar technology. Earlier in 2008, India launched Israel's spy satellite TecSar from India which was hailed as an important milestone symbolizing the growing military ties between the two countries. Later in Apr 2009, India launched Israeli designed an all-weather surveillance satellite RISAT-2 for the security establishments. The satellite is known to give India the ability to monitor cross-border movements of suspected terrorists, as well as troop movements in Pakistan and other neighbouring countries, at night and under all weather conditions.[93] India has also become the largest client of Israel's defence industry, a largest supplier of defence equipment after Russia. Focus in Defence ties is on joint research and development projects. Agreements on joint development of Surface-to-Air Missiles (SAM) and interceptors, ship-launched rotary unmanned aerial vehicles, and advanced radars are indications that Indo-Israeli defence cooperation is moving from more traditional arms deals to long-term cooperation. Already India has imported and modified Israeli technology in missile defence. One example is the Long-Range Tracking Radar (LRTR) which was used in the "exo-atmospheric" BMD system. Co-operation also demonstrates that India is able to take advantage of niche technologies developed outside, and incorporating them into indigenous systems. Israeli companies have also become very active in modernizing Soviet/Russian military equipment, which is beneficial to India.

Russia. Russia and India gradually moved from a supplier-client relationship to one where joint development and marketing of weapons systems is emphasized, while protecting intellectual property rights of Russian technology through an IPR agreement. GLONASS navigation system, Fifth Generation Fighter Aircraft (FGFA) and the BrahMos supersonic cruise missile are examples of joint projects with Russia. Like India's co-operation with Israel, these agreements enable India to develop advanced technology and weaponry without necessarily having indigenous

---

[93] *India launches a $ 200m TECSTAR spy satellite,* available at http://www.defenseindustrydaily.com/Indias-200M-TECSAR-Satellite-Purchase-Launched-05386/

capability, as well as reduce reliance on any one outside actor.

In December 2004, recognizing the need for reliable Navigation and targeting system and to collect data on launches of mobile ballistic missiles of the adversary, India reached an agreement with Russia to jointly develop the Russian GLONASS global navigation satellite system. Under the agreement, India will become the sole partner of Russia on GLONASS development, replenishment and future use. It is an alternative and complementary to the United States' Global Positioning System (GPS), the Chinese Compass navigation system, and the planned Galileo positioning system of the European Union (EU). After America's Global Positioning System (GPS), GLONASS is currently the only other satellite-based navigation system and it would be available for Indian military applications. The Sukhoi Su-30MKI multi-role fighters supplied to the Indian Air Force are already equipped with GPS and Global Navigation Satellite System (GLONASS) receivers for navigation. America may, however, deny use of this during operations or may introduce errors to render it inoperable; the US is also in talks with the European Union to ensure that the Galileo system is not made available for military applications. India's goal to be a regional military power is dependent on access to satellite navigation. The constellation of GLONASS satellites will be available for use by civilian and military users in both India and Russia. It is expected to be restored to fully deployed status (i.e. 24 satellites in orbit and continuous global coverage) by 2011. The GPS's potential in contributing to precision warfare has been tested in some parts; particularly in the Iraq war by the U.S., and widely discussed globally. With India's global position in the subcontinent, as well as its problems with terrorist movements, access to such a system is expected to give both its military and civilian administration a clear advantage over its potential foes.ISRO has also planned to build an independent satellite navigation system using home-grown components. The project, called the Indian Regional Navigation System, will be implemented by 2014 at the cost of Rs.1,600 crore.It will consist of a constellation of seven or eight satellites, as well as a large ground segment.[94]

United States. In June 2005, the United States and India signed a 10-year defence pact guiding strategic co-operation on defence collaboration in the areas of two-way defence trade, technology transfers and co-production, expanded missile defence collaboration, and a bilateral defence procurement and production.[95] Since 2002, there has been talk of expanding co-operation on missile defence and other advanced technologies. The United States and India continue to explore cooperation on missile defence, and even a joint missile defence shield. Secretary of Defense Robert Gates indicated in late February 2008 that the United States is seeking a steady expansion of the defence relationship with India.

Analysis of the technology and strategic priority indicate that in time to come, space would get weaponised perhaps in the phased manner. Despite overt opposition by Russia and China who in year 2002 had put up a joint proposal in UN conference on Disarmament (CD) for prevention of Arms race in the outer space (PAROS), the possession of ASAT weapons by Russia and recent ASAT test by China in destroying its own weather satellite in Jan 2007, raises question on their commitment in prevention of weaponisation of the outer space. In fact, it appears to be a tactical ploy of these nations to gain more time for gaining some parity with US. Under the circumstance, it is imperative for the space faring nations like India step up R &D in the space technologies and exploit the space for military besides, other civil uses. The development of critical technologies such as missile defences, advanced sensors, miniaturization techniques, high power lasers etc must be pursued for counter space technologies. The prime objective should be to bridge the technology gap with advanced nations so that India does not become one of the targets for the space arms control.

---

94 en.wikipedia.org/wiki/Indian_Regional_Navigational_Satellite_System

95 http://www.defenseindustrydaily.com/us-india-sign-10year-defense-pact-0783/

THREATS TO INDIA'S SPACE ASSETS

Ever since India gained independence from the colonial rule, Pakistan has remained central to the India's national security consciousness. After 1962 India china war, China emerged as Pakistan's single most trusted and enduring military ally. Introduction of China in the nexus has changed the dimension of threat, though many now believe that India-China relationships have become so multifaceted, and economic ties have bloomed and reached $50 billion annually (Chinese premier, Wen Jiabao during his visit of India in Dec 10 assured its growth further to US$100 billion by 2015) that direct military confrontation with China is a remote possibility. But, in background of India-China War of 1962, unresolved border issues and persistent Chinese claim on Arunachal Pradesh, relying on such a belief would be a folly as in the past within two months of the conclusion of the of the Sino-Indian agreement on trade, first Chinese intrusion took place on 17 July 1954 in the Uttar Pradesh Sector. Hence, when clash of interests takes place irrespective of trade ties, a possibility of war can not be ruled out. Strategists believe that cause of 1962 war was mainly due the trust deficit. China then perceived real threat from Taiwan and South Korea with support of United States. To them India was a virtual colony under the domination of US imperialists and regarded India as a conglomeration of nationalities. China's war with India was meant to pre-empt any simultaneous US Indian intervention from its East which in reality was not to be, as India professed non-alignment in the Nehru era and was never ever militarily aligned with America and has itself no territorial ambitions.[96]

---

[96] USI Paper titled " *The Real Story of China's War on India* ,1962(2006) By Shri AK Dave IPS(Retd).

China still perceives American threat and is striving to match it at the strategic and conventional level. On the other hand, America's cold war competitor of yester years, Russia yearns to regain its lost superpower status. America possesses considerable lead in space supported net centric warfare with its superior technology. Further, America unilaterally withdrew from bilateral Anti Ballistic missile treaty in 2001 by citing its irrelevance and terrorism threat. The American announcement came after failing to persuade Russia to set the treaty aside and negotiate a new strategic agreement. President Bush advocated scrapping the ABM treaty, calling it a relic from a much different time," as the 1972 ABM treaty was signed by the United States and the Soviet Union at a much different time, in a cold war era and one of the signatories, the Soviet Union, no longer exists and neither does the hostility that once led both countries to keep thousands of nuclear weapons on hair-trigger alert, pointed at each other.[97] Both China and Russia are opposed to the U.S. Missile Defence Programme fearing that it could lead to dilution of their nuclear deterrence and thus seek another multilateral agreement to stop America from pursuing its plan. This perception had probably prompted both Russian and Chinese Presidents to announce on 17 June 2010, after the Ninth Heads of State summit of the Shanghai Cooperation Organization (SCO) in Yekaterinburg, Russia, that "Russia and China advocate peaceful uses of outer space and oppose the prospect of it being turned into a new area for deploying weapons" and they are drafting a joint treaty to be presented to the United Nations General Assembly to ban the deployment of weapons in outer space. Apparently, the statement by the Presidents reflected a common purpose to avoid the militarization of space but, it also brought out their concern on lagging behind the United States in the space technologies. Both countries would continue to raise issue of militarization of the space till they achieve some form of parity in space technology, despite the fact that all three states have the Anti Satellite capability.

---

[97] U.S. quits ABM treaty, CNN News, December 14, 2001, http://archives.cnn.com/2001/ALLPOLITICS/12/13/rec.bush.abm/

The US has consistently opposed their efforts to seek a multilateral agreement or treaty which could block their research in the futuristic area of space. The true reason behind the American desire for global anti-ballistic missile defence and space militarization is that the United States believes that over the next two to three decades, it can beat the others (Russia and China) in these spheres and gain a decisive strategic military advantage. US Air Chief was quoted in an Air Force report entitled "Spherical battle space is new theatre of operation" as saying" I think for far too long we have looked at our conception of future battle space by standing on the ground and looking up - I think that might be the wrong way to look."

According to the Chinese, the United States and Russia are engaged in a race to develop ground, air, and space-based weapons for achieving space dominance. These are said to include ground-based kinetic and airborne ASAT systems, high-altitude anti-missile weapons, space weapons platforms, aerospace aircraft, and space combat aircraft designed to execute simultaneous space and ground strikes.

Comparative Military Balance

China and India as rising Asian powers with high GDP growths and increasing geo-political influence are in race for regional dominance. China has officially acknowledged defence spending of about 78 billion US dollars in 2010. A spokesman for the annual session of the National People's Congress (NPC) claimed that China's defence expenditure accounted for about 1.4 percent of its GDP in comparison to four percent for the United States, and more than two percent for the United Kingdom, France and Russia.[98] However, this assertion may not be true, as the actual Chinese military capabilities and budget are shrouded in deep secrecy perhaps to create the strategic advantage of uncertainty and to prevent other countries

---

[98] China Defence Budget to grow 7.5%, a report in *China' Daily,* 04 Mar 2010 http://www.chinadaily.com.cn/china/2010-03/04/content_9537753.htm

having an idea of its military might. Its actual defence spending is could be between 150%-300% higher than revealed I.e, some where in the region of $150billion (which would be closer to 3% of their GDP). Even If we were to go by the conservative official Chinese figure of 78 billion, it would put China second only to USA in global military spending. On the other side, India's defence budget for 2010 is quoted at $31.9 billion (Rs 1,47,344 crore) i.e. 2.12 of GDP. However, unlike China, India does not keep a cloak of secrecy, as its democratic government system requires public accountability.

A comparative analysis shows that in conventional military strengths, China is superior to India. India's active military strength is over 1,325,000 while China's military personnel strength is significantly higher at 2,255,000. In air defence, China's People's Liberation Army (PLA) Air Force has 9,218 aircrafts of which about 2300 are combat aircrafts, operating from its approx 230 air bases. The Indian Air Force has 1700 aircrafts which includes 852 combat aircrafts operating from its approx 60 bases and its sole aircraft carrier INS Viraat. The air superiority in China's PLAAF is maintained by its fleet of Russian Su-30 MK and indigenously built J-10 fighters. Indian Air Force, on the other hand has French built Dassault Mirage 2000s and Russian Su-30 MKI as the best aircraft in its combat fleet. Indian Navy is the world's eighth largest Navy with a fleet of 145 vessels consisting of missile-capable warships, advanced submarines, the latest naval aircrafts and an aircraft carrier in its inventory. It is experienced both in combat and rescue operations during wartime and peace as seen from its wars with Pakistan in 1971, the December 2004 Tsunami, etc. In comparison, China's PLA Navy with its fleet of 284 vessels and a force of more than 60 submarines is quantitatively larger but lacking in actual war experience. Moreover, it lags behind Western standards as most weapons and sensor systems are based on older Russian technologies, which could undermine its strategic capability. As of now, China has no aircraft carriers in its naval fleet but was slated to retrofit and induct an aircraft carrier by 2010 which now is reported to be functional.

In strategic nuclear defence and delivery systems, China's PLA is miles ahead of India's nuclear forces. The PLA's stockpile is estimated to have 200-400 active nuclear warheads. In comparison, India's strategic nuclear force is estimated to have stockpiled about 50-70 nuclear warheads. The most powerful warhead tested by India had an yield of 20-25 kiloton which is quite small compared to China's highest yield of 4 megatons. India's nuclear delivery system consists of bombers, supersonic cruise missiles and medium range ballistic missiles. Agni II, India's longest range deployed ballistic missile is capable of a range of 2500 km, carrying a single nuclear warhead of ~1000 kg. (Agni III is still not operationally deployed) even after induction of Agni-III. India can only hit a limited number of Chinese targets. Agni-V with the range of 4500-5000km which is still to be test launched would cover targets in Eastern China. India is also developing the Surya ICBM, which is based on civil space launch technologies from India's Space Launch Vehicle Programme. Once completed, the Surya will extend India's nuclear deterrent to targets deep within China and will thus provide India with a strong deterrent against future Chinese aggression. In stark contrast, China's nuclear delivery system is far more capable with multiple warheads (MIRV) ICBMs like DF-5A [12000+ km] and DF-4 [7500+ km]. It also fields submarine launched SLBMs like JL-2 [8500+ km] and strategic fighter bombers like Su-27 Flanker in its nuclear delivery arsenal.

China, perceives a powerful military adversary in United States. A possibility of conflict with US over Taiwan, gives China a strong incentive to beef up its military defence to counter the US military might. The situation is much similar to that of USSR vs USA Cold War, although on a much smaller scale. China's massive military modernization, vigorous efforts to develop a range of space weapons and heavy-lift space vehicles, and a sustained move towards increasing the range and lethality of missiles are not merely exercises to compete militarily with the US. Their purpose is also to deter American intervention, should Beijing decide to overrun Taiwan by force and project itself as the undisputed leader of Asian region. Chinese are aware that post 1962 India is no more a push over, but are not unduly concerned as India lies well within the lethal ranges of their weapons while the opposite is not true as yet.

India being a democracy, though perceiving Chinese threat, its military modernisation drive is often overshadowed by internal militancy issues and political struggles. The end result is China is far ahead of India in military might with overpowering superiority if both conventional and nuclear forces are taken into account.[99]

China has seemingly have resolved all border disputes with all its neighbours except India. Chinese People's Liberation Army (PLA) is concerned with strategic ramifications of India's emergence as an economic, military and political power. Even as China shows keenness to improve its ties with India, its military incursions in Sikkim, Ladhakh and the line of actual control (LAC) in Arunachal Pradesh, and its growing defence ties with Pakistan, remain matters of concern for India.

Chinese Space Threat

Space systems are, by their very nature, vulnerable to a range of threats. These threats include jamming satellite links or blinding satellite sensors, which can be disruptive or can temporarily deny access to space-derived products. Anti-satellite weapons whether kinetic or conventional or Electro-Magnetic Pulse (EMP) weapons can permanently and irreversibly destroy satellites. Military force can be employed against ground relay stations, communication nodes, or satellite command and control systems to render space assets useless over an extended period of time. Adversaries can also employ denial and deception techniques to confuse or complicate our information collection.

Despite overt opposition to the weaponisation of the space, Chinese believe that the weaponisation of space is an inevitable developmental

[99]India vs China on military strenght:Conventional and Nuclear, http://www.abytheliberal.com/internationalism/india-vs-china-military-conventional-nuclear.

trend and the strategic competition in the 21st century will not be on Earth, but in space. Realising this, for more than a decade, Chinese military strategists and aerospace scientists have been quietly designing a blueprint for achieving space dominance through military modernisation. The prime reason for the accelerated rate of military modernisation has been due to a fundamental restructuring of the Chinese defence industry. The restructuring has resulted in an accelerated rate of military system modernization especially in defence electronics.[100]

Against this backdrop, the prospects of PLA's swift emergence as a challenger in space are now almost certain. The other indication is that the Chinese decided to cancel weapons projects that had been active for 10 years or longer and to direct these funds to developing so-called "new-concept weapons" 'viz; laser, beam, electromagnetic, microwave, infrasonic, climatic, genetic, biotechnological, and nano technological. China now has three military priorities: space, nuclear weapons, and "new-concept" weapons.

Chinese military scientists define space warfare as combat operations whose major goal is to seize and maintain space dominance, whose major combat arena is outer space, and whose major combat strength is military space forces. Future space warfare would include dogfights between the space-based combat systems of belligerents; intercepts of strategic ballistic missiles by space based combat platforms; strikes by space weapons on

---

[100] China's approach to civil-military integration began to change around the mid-1990s, and it entailed a crucial shift in policy, from conversion (i.e., switching military factories over to civilian use) to the promotion of integrated dual-use industrial systems capable of developing and manufacturing both defence and military goods. This new strategy was embodied and made a priority in the defence industry's five-year plan for 2001- 2005, which emphasized the dual importance of both the transfer of military technologies to commercial use and the transfer of commercial technologies to military use, and which therefore called for the Chinese arms Industry to not only to develop dual-use technologies but to actively promote joint civil-military technology cooperation. The key areas of China's new focus on dual-use technology development and subsequent spin-on include microelectronics, space systems, new materials (such as composites and alloys), propulsion, missiles, computer- aided manufacturing, and particularly information technologies.Asia pacific centre for security studies volume 3 number 9,Dec 2004 Http://www.apcss.org/Publications/APSSS/Civil-MilitaryIntegration.pdf

Earth targets and Earth-based counter space or space defence operations; and strikes from the land, sea, and air on enemy space launch platforms and command-and-control organs. Chinese military scientists contend that space warfare will become the core of future non-contact combat. The integrated space-based combat platforms, weaponry, and C4ISR components will guide the various combat elements of the three armed services to launch long-distance precision attacks on ground, sea, air, and space targets.

Current Status

On 11 Jan 2007, China successfully destroyed an old weather Satellite deployed in a low Earth orbit at an altitude of 864 kilometres with a ground launched missile. A year before that it had dazzled a US reconnaissance Satellite using a ground laser. Although it is often argued that China's anti satellite weapon test was a protest against U.S. space policies. But the fact is, it was part of well considered China's Space denial strategy to achieve dominance.

Direct Attack Weapons. The shooting down of the US satellite was the example of the direct attack weapon. These are particularly effective against satellites flying in low Earth orbits, where most of remote sensing, meteorological, and imaging (electro-optical, infrared, and radar) intelligence satellites and currently operate. They can also threaten spacecraft in medium and geosynchronous orbits, however, provided the attacker has a sufficiently powerful booster. This is where navigation and guidance satellites, military communications platforms, and early-warning and nuclear detonation systems now operate. There are several Chinese space launch vehicles and ballistic missiles like the DF-31 that could easily carry an ASAT payload to geosynchronous orbit, and it is not unreasonable to expect such dedicated systems in the future.[101] *The direct attack*

---

[101] Punching the U.S. Military's "Soft Ribs":China's Antisatellite Weapon Test in Strategic Perspective By Ashley J. Tellis, Senior Associate, Carnegie Endowment for International Peacehttp://www.carnegieendowment.org/files/pb_51_tellis_final.pdf

however, leads to increase debris will pose enormous risk to all the other satellites that exist in the orbit. *The Chinese* test appears to have increased the amount of debris (size greater than 1 centimetre) in Low Earth Orbit (LEO) by 15 to 20 per cent, becoming the worst debris-producing event on record.[102]

Directed-Energy Weapons. As part of the effort to develop "new concept weapons," China has devoted substantial resources to directed energy systems, particularly ground-based high- and low-power lasers, for counter space purposes. Other technologies being discussed in China include high-power microwave weapons, electromagnetic rail guns, and particle beam systems. Lasers are particularly attractive counter space weapons because cause varying levels of satellite damage depending on necessity. A low-power laser, for example, could be used to temporarily blind an electro-optical intelligence collector by over saturating its receptors. A high-power laser could be used to actually inflict structural damage on a spacecraft. Satellites in any orbit can be attacked by ground based lasers; assuming that beam quality, jitter and control, and propagation problems have been satisfactorily resolved; the vulnerability of high-altitude satellites would depend mainly on the power output of the laser. China is already known to have used laser on U.S. reconnaissance satellites, and its capability to inflict more consequential damage will only grow over the next decade.[103]

Electronic Attack. When "hard kill" in space appears beyond reach of ground attack weapons because of the greater distances of satellite's orbit from Earth, Chinese military planners believe that sophisticated jamming technologies that could enforce information blackouts at critical moments

---

[102] Does India really need ASAT capability now? by Bharath Gopalaswamy and Harsh V. Pant.http://www.defence.pk/forums/military-forum/44983-india-targets-chinas-satellites-3.html

[103] Punching the U.S. Military's "Soft Ribs":China's Antisatellite Weapon Test in Strategic Perspective By Ashley J. Tellis, Senior Associate, Carnegie Endowment for International Peace, http://www.carnegieendowment.org/files/pb_51_tellis_final.pdf

in a war. Electronic attack would thus be carried out to disable space assets located in medium, geosynchronous, and highly elliptical orbits for a specific period to deny the useful intelligence. The most important of these are the military tactical communications platforms and the global positioning system constellation, which provides precision navigation and timing data to military users and permits the accurate targeting of various weapon systems.[104]

Orbital Platform. In 2003, China became the third country in the world to successfully hoist an astronaut into space. In 2005, China repeated this feat by putting two astronauts into space. This was followed by a successful space walk in September 2008. The successful completion of the human space flight and space walk has given impetus to China's plan to build and launch an orbital complex. An orbital complex, besides helping China undertake cutting edge research activities, could serve in future as a strategic platform in space to bolster China's space war efforts. The Chinese plan for a permanent space station would give it an edge in the event of a war involving space assets. China could use its space complex to either shoot down a missile or kill a satellite. Space dominance, however, like the one the US military presently enjoys still remains a distant dream for China's military.

Ground Infrastructure. China also considers employing means to make the launch site inoperative through hard and soft kill . This makes our only Satellite Launch site at Satish Dhawan Space centre, Sriharikota particularly vulnerable to damage/destruction through sabotage. Compared to India, China has at least three space launch sites Whereas, United States and Russia have several sites which gives a relative security from intentional hostile action. Therefore, apart from foolproof security, we also need to replicate the facility at other places to retain our capability to launch satellites on demand if necessary during war. Chinese may also be capable establishing a space monitoring system to acquire the satellite's technical

[104] Ibid

parameters and character information, and effectively detect and analyse the adversary's satellite's operational system and down-link remote signal and attack the satellite's space-to-ground communication and command nodes to weaken the connection, link, mutual operation, and networking flexibility in order to degrade its operational effectiveness of the satellite to the adversary.

Notwithstanding, China would need to achieve a greater degree of sophistication to destroy enemy satellites meant for end-uses, such as communications, surveillance and navigation. For such satellites are placed into a higher orbit. But it is easier to kill reconnaissance satellites, a majority of which move in low earth orbits. Imaging surveillance satellite would however would be vulnerable to blinding by China's laser beams because their functions depend on devices which are sensitive to light.

Currently, India has 12 satellites in LEO out of which, RISAT-1 is probably an attractive target. RISAT-1 is an experimental satellite, which has the capability to operate in all weather conditions. RISAT-1 is believed to be dedicated for military applications. On the other hand, China has 31 satellites in the LEO orbit, out of which 12 of them are dedicated for military purposes. A space war (mutual shooting down of satellites) between China and India will be devastating. India's lack of redundancy in satellite capabilities will compromise its capability to retaliate. The resultant debris will pose enormous risk to not only Indian and Chinese satellites but also to all the other satellites that exist in the orbit. Thus mutual shooting down capability as well as the debris would deter use of direct attack weapons.

Pakistan's Space Threat

Pakistan Space and Upper Atmosphere Research Commission (SUPARCO) was established in 1961 to meet the space aspiration of Pakistan. Its first satellite Badr-1 (Badr-A) from the Xichang Launch Centre, China three decades later on July 16, 1990 aboard a Chinese Long March 2E rocket. Second satellite BADR-B an Earth Observation Satellite designed by Space Innovations Limited of UK followed after another after another decade on 10 December 2001 which was launched on a Zenit 2 rocket from Baikonur Cosmodrome, Kazakhstan. Paksat 1 Pakistan's first geostationary satellite manufactured by Boeing for Indonesia and launched on February 1, 1996 was leased to Pakistan in 2002. Paksat-1R satellite will replace the existing Paksat-1 in 2011.[105] Pakistan has followed a Policy of deliberate ambiguity for many decades. This is why it is still unclear what the plans and operation as well as capabilities of SUPARCO and its space facilities are. Despite SUPARCO civilian orientation, the agency is believed to be involved in development of short and medium range solid fuelled ballistic missiles for Pakistan's military. Pakistan has indigenously manufactured Hatf-1 a short range solid fuel missile with a range of about 100 km and pay load of 500kg and Hatf-2 with a similar pay load up to the range of 300 kg. However, analysts believe that these missiles are based on French sounding rockets Technology.[106] The other missiles namely Shaheen I (range 700 km), Shaheen II (IRBM range 2000km), Ghauri, Babar, Abdali are more likely copies of Chinese missiles.

On March 2001, Dr. Abdul Qadeer Khan announced that Pakistani scientists were in the process of building the country's first Satellite Launch Vehicle (SLV) and that the project had been assigned to SUPARCO, which also built the Badr satellites Dr. Abdul Qadeer Khan also added that "Pakistan has very robust IRBMs which can launch geostationary orbiting satellites. All Pakistan has to do is to erase Delhi or Kolkota from the target and point it towards the sky. Instead of Hydrogen bombs and Atomic bombs the missiles can easily carry a payload of a satellite".

Unlike ISRO which is essentially a civilian setup SUPARCO on the contrary is headed and administered by Pakistan military. Pakistan has ambition in respect of space but it lacks the capability and wherewithal at

---

[105] Pakistan Space Programme, http://www.aerospaceguide.net/worldspace/pakistan.html

[106] http://www.aerospaceguide.net/worldspace/pakistan.html Space the frontier of Modern Defence By SQn Ldr K K Nair page no 136 , 2006 edition.

present. But, considering its closeness with Chinese, it be get the required support from the Chinese. However, in view of the financial constraints, Pakistan may not be able to achieve the space capabilities and does not pose a very potent direct attack threat to our satellites as yet but, our single space launch facility remains prone to sabotage and physical damage.

INDIA'S MILITARY OPTIONS IN SPACE

India has a strong civilian space programme but, the use of space or space-related technologies for military purposes has been rather limited so far. For a nation which has fought over four major wars since its birth and a variety of insurgents, can India continue to remain committed to the existing policy even as counter space systems are emerging in our neighbourhood and as the ongoing developments in the military use of satellites provide a pointer to the future threat to the Indian space assets.

Till recently, India had desisted from launching dedicated Military satellites despite possessing the technical wherewithal and being one of the leading spaces faring nation. However, after the Chinese anti satellite test, the issue on space security has received more attention in defence, media and public debates. Taking cognizance of this, Ministry of Defence established a tri-services space cell with in Integrated Services Headquarters in 2008. The Space cell while acting as single window agency to help Indian Defence forces to gain an access to the constellation of satellites operated by the Indian Space Research Organisation (ISRO), will also serve as the forerunner of a full-fledged aerospace command, whose requirement was recommended by Indian Parliamentary Committee on Defence in 2000 and later reiterated by the Standing Committee on Defence but formation has been deferred for a long time now.[107]

Indeed, the Chinese ASAT lethality arguably holds the greatest import

---

[107] Standing Committee On Defence (2003), (Thirteenth Lok Sabha) Nineteenth Report Ministry Of Defence, Http://164.100.24.208/Ls/Committeer/Defence/19.Pdf

for India. The only counter to ASAT weapons is a capability to pay back in kind. Both United States and Russia can cripple China's communications and expose its ground assets if their space assets are ever struck. Japan, also concerned over the test, is protected under the U.S. security umbrella. India, by contrast, neither has the missile reach for a counter-offensive in the Chinese heartland nor has ASAT power to deter the destruction of its space assets.

India's Options

The end of last millennium witnessed two major conflicts in the world, one in the Serbian province of Kosovo and another on the icy heights of Kargil. In Kosovo, considerable use of space assets was made during the conflict. As many as 50 satellites were brought to bear upon a single war. In contrast, the Kargil conflict, not a single spacecraft was used despite that India was an established Space power by then. Had India harnessed the space capabilities for its national security in 1999, our forces could have detected the enemy excursion and assessed the scale of excursion more accurately, our forces would have delivered fire power and applied the air power precisely and persistently and finally we would have suffered less in terms of human lives and limbs therefore, induction and integration of space capabilities into conventional military a capability is inescapable.

Outer Space Policy

Space technologies have the global reach beyond the national borders and, hence, a suitable policy and organizational infrastructure are vital to bring together all components of space power for deterrence, war fighting and or for power projections to support the national interests. It is likely that in the time to come the deterrence value of space technology may become as important as the nuclear deterrence is today, the space will then be an important new dimension for calculating for military and economic power of the nation. India therefore, needs a well calibrated 'Outer Space Policy' that would not only enhance the civilian space profile but, also enable development of suitable counter space capabilities to protect its own security concerns. Such an integrated approach alone will enable India to claim the rightful place in the comity of nations. Realizing this, Ministry of Defence has tasked HQ IDS to formulate the Defence Space Vision-2020, outlining the roadmap for the armed forces in the realm of space, with intelligence, reconnaissance, surveillance and navigation as the thrust areas in its first phase[108].

Indian armed forces have been already using "dual use" satellites (Cartosat-I, Cartosat-II series of satellites ,RISAT-2 TES etc). They will get dedicated satellites of their own under Defence Space Vision 2020. First dedicated military satellite for Navy, with an around 1,000 nautical mile foot-Print over Indian Ocean is slated to be launched in 2011. IAF and Army satellites are also expected to follow in a couple of years. However, the Armed forces are still distance away from exploiting space for real-time military communications and reconnaissance missions; leave alone for uses like missile early-warning, delivery of precision-guided munitions through satellite signals or jamming enemy networks.[109]

Separate Space Command

Major space faring nations all over the world have dedicated space organizations to look after the defence and civil satellite needs. In United States, the Military and civil space programme are separated organizationally. The space activities are coordinated by the Department of Defence which is equivalent to our Ministry of Defence (MoD) however, civil and military use the same launch vehicles and launch facilities. A separate Space Command controlled the military space activities which in 2002 have been merged with the US Strategic Command for achieving better operational efficiency. The merged command operates the Space

---

[108] Times of India   06 Feb. 2007 http://timesofindia.indiatimes.com/NEWS/India/India_years_away_from_setting_up_aerospace_command/articleshow/1565014.cms, http://mod.nic.in/aboutus/welcome.html

[109] Ibid

Defense Operations Center (SPADOC), the Space Surveillance Center (SSC), the Missile Warning Center (MWC), and the Joint Space Intelligence Center (JSIC). It provides strategic and theatre ballistic missile warnings to US nation's leadership and to deployed troops world wide, constant global connectivity with deployed forces, precise navigation, targeting and timing support to coordinate the positioning and manoeuvre of U.S and allied aircrews, naval forces and ground forces, collects and distributes global weather data and coordinate space-based imagery between intelligence agencies and planners within Unified Commands.[110] The civilian space programme and aeronautic and space research is looked after by National Aeronautics and Space Administration (NASA). Similarly, the military and civil space programme was separated post break-up of the erstwhile Soviet Union. The Strategic Rocket Forces control the Russian military space activities whereas; the Russian Space agency (RKA) looks after Russian space science programme and general aerospace research. Like wise, French civilian and military space programme are also separated. Space Coordination Group, headed by the Chief Officer of Defence, oversees the collaboration between the Defence Procurement Agency, which manages military space programs, and the civilian national space agency, National Space Studies Center (CNES), which is in charge of developing and operating space systems. Overall authority for military space policy falls under the Defence Ministry.[111] As per the French white paper on defence, a dedicated Joint Space Command, under the authority of the Chief of Defence Staff will be established to control the military space functions.[112] As for China, its National Space Administration (CNSA) is responsible for the management of space activities for civilian use and international space cooperation with other countries, whereas, all military

---

[110] http://www.fas.org/spp/military/program/nssrm/initiatives/usspace.htm

[111] http://75.125.200.178/~admin23/index.php?id=105&page=France_Military

[112] *THE STRATEGIC USE OF OUTER SPACE, The French White Paper on defence and nationalsecurity,* http://www.globalsecurity.org/military/library/report/2008/livre-blanc_france_2008-14.htm

space operations are controlled by People Liberation Army (PLA).[113],[114] As regards to United Kingdom, it has reliance on United states for security and defence technology. For this reason it has no defence satellites of its own. UK's main strength is in telecommunications. Assistant Chief of Air Staff of RAF coordinates the military activities across the MoD,[115] whereas, UK Space Agency (erstwhile British National Space Centre) coordinates the civil space activities in United Kingdom.[116] In India, earlier there was virtually no dedicated body to look after the space needs of the defence forces. In 2008, Integrated Space Cell was established after the security re appraisal in wake of the Chinese ASAT. In the present form and skill level, it can at best can project the space requirements of the forces to ISRO but, cannot optimally integrate air and space-based assets and support the offensive and defensive operations in real-time to realize its impact as a force multiplier. The space forces tasks would be mainly centered around surveillance, intelligence, communications, navigation and accurate weapon delivery and damage assessment. These operations can not be coordinated by the space cell and would require a full-fledged tri-service command manned by the trained staff to meet the requirements of three services. US space command for example is manned by around 25400 trained civil and military personnel. The current space cell can function as a forerunner to the future tri-service space command as and when established and provide a training ground to meet the huge requirement of the trained staff. On formation of the dedicated space command, our neighbours may raise hackles alleging fuelling of the arms race in the region, however, India is not the initiator of military use of the space in this region. We are forced to look at this option only because of heightened security environment created by our neighbours. Any further delay in

---

[113] White paper on China 's space activities 2006, http://www.spacedaily.com/reports/China_Issues_White_Paper_On_Space_Activities_999.html

[114] http://en.wikipedia.org/wiki/People's_Liberation_Army

[115] Parliamentary Office Of Science and Technology ,POSTNOTE December 2006,number273 – MILITRAY USES OF SPACE http://www.parliament.uk/documents/post/postpn273.pdf

[116] http://en.wikipedia.org/wiki/British_National_Space_Centre

formation of the space command could be only be at the cost of operational efficiency that space assets would bring. Being a tri service institution it should be under the HQ IDS with functional linkage with DoS/ISRO which besides pursuing its own civilian programme should design satellites and provide launch facilities to defence.

Missile Defence System

Dependence on space based assets make them vulnerable to attack by the enemy ,hence, to retain the war fighting ability we must embark upon a dedicated initiative to save our satellites assets. The missile defence system currently being developed by defence research development Organisation (DRDO) for exo-atmospheric and endo-atmospheric engagement of Ballistic missiles is restricted to engage ballistic missile threat up to range of 600 km despite the fact that the Interceptor missile is stated to be capable of engaging of missile threats up to the range of 2000km. Satellite assisted detection can help in realizing full system potential. DRDO director has publicly stated that based on the experience gained in the BMD, DRDO could develop anti satellite weapon if needed. The mutual assured destruction capability of satellites will provide security to own satellites just in the same manner as for nuclear deterrence.

Network Centric Warfare (NCW). The NCW is the product of the IT revolution. It envisages the integration of information from all sensors and making it available as required, wherever needed by the authorized recipients. NCW facilitates information sharing across multiple levels of traditional echelons of command and control through satellite links. All the three services—army, navy, air-force—should be integrated by linking all the radars and sensors through the satellite system for early warning and control system. The object is to provide very high level of the situational awareness that will in its wake lead to greater efficiency in the prosecution of the war. The role of satellites lies in reconnaissance of the troop movement of the enemy, for tapping communication, jamming the enemy network and providing real-communication from commander right up to the troop level within the constraints of our resources. In a conflict environment where both adversaries have the NCW capability, before gaining the advantage of own NCW it is necessary to make the enemy NCW non-functional by jamming it or damaging the NCW assets.

Drone Technology. Another emerging area is Drone Technology which does not have the desired efficacy and coverage without satellite links. The effectiveness of the drone in surveillance and in attack roles can be seen in military operations in Iraq war and in Afghanistan spearheaded by US and allied forces. The entire range of reconnaissance information systems, including tactical UAV (Unmanned Aerial Vehicle) imageries, were analyzed at a central facility and transmitted back to the ground forces with minimum time-loss through a satellite communications system.

Satellites in Nuclear Environment

India has declared "**No First Use**" of Nuclear weapons and would exercise the option of massive second strike if attacked. This however is possible only if we have capability to monitor the launch of enemy ballistic missiles for which thermal imaging satellites are a must. The satellite detection would enable launch of counter measures to neutralize the threat .This requirement is considered inescapable due to the availability of short reaction time from the missile threat of our potential adversaries.

Global Positioning System

The recent wars, there was extensive use of Global positioning Systems (GPS) not only for all weather navigation but for accurate targeting as well. Aerospace weapons like **Joint Direct Attack Munition (JDAM)** can heavily influence the war as the American experiences in Iraq and Afghanistan have shown. In fact, they are futuristic weapons are guided to their target by an integrated inertial guidance system coupled with a Global Positioning System (GPS) receiver for enhanced accuracy longer range. We can not rely on the availability of American NAVSTAR GPS during hostilities. Though, we being a partner in Russian GLONASS navigation system and its availability is assured yet, we have plan to build an independent navigation faculty a constellation of seven satellites, the Indian regional Navigation Satellite system (IRNSS) which will start functioning by 2014 and can play crucial military roles besides other civil tasks.

Defending the Space Assets

Twenty one Indian satellites are currently in operation. Of these ten remote sensing satellite(IRS) are in the Low Earth Orbit (LEO) are facing the potential threat of damage and destruction from the direct ascent anti satellite weapons The number of nations with the ability to develop weapons capable of attacking space-based assets, at least those in LEO, is increasing steadily. Direct ascent anti-satellite weapons are well within the capabilities of any nation able to place a satellite in orbit, and possibly some capable of building only sounding rockets. The Geostationary satellite and satellites at the medium earth orbit (MEO) are presently well above the effective envelop of the kinetic weapons but these are susceptible electronic attacks of the enemy. Electronic attacks will generally take two forms i.e. uplink jamming and downlink jamming. All military and commercial satellite communication system is susceptible in varying degrees to both types of jamming. Uplink jamming targets the radio receiver component of the transponder, including sensors and command receivers and it usually requires high powers transmitters. The targets of downlink jamming typically are ground based receivers which are easy to jam and require relative less power. During the operation of Iraqi Freedom Saddam Hussein's forces employed Russian GPS jammers against the coalition forces. Although the attempt of jamming did not succeed it speaks volumes of the means that adversary could employ.[117]

Protecting Launch Facilities. Maintaining access to space requires that all parts of a complex system be protected. The main components of this system are the satellites themselves, the ground stations that control

---

[117] Space Technology and Network centric Warfare: A Strategic Paradox USAWC Stragec Research Paper 2006

them, and any mobile ground stations capable of receiving satellite generated data. Probably the most vulnerable part of the system is the network of ground stations. Given the fact that they are relatively few in number and are "soft" targets with known locations, attacks against ground stations should be expected in the event of war. It is possible to deploy mobile ground stations, using them will almost certainly degrade the efficiency of the system. Equally vulnerable are the facilities needed to launch satellites into orbit. Although they are somewhat more difficult to destroy, the need to defend launch facilities is potentially greater since they are fewer in number and making them mobile is extremely difficult. Multiple dispersed facilities would provide redundancy in case one facility is rendered non operational.

Protection of the facilities is more a matter for conventional security arrangements than protection of space-based assets. However, it is important to note that any investments made in space-based weapons would be wasted if the ground stations needed to make use of them were neutralized.[118]

Protection Through Redundancy. Until very recently, satellites have tended to be fairly large, very capable and very expensive. These satellites present an opponent with lucrative targets, where the loss of even one would often constitute a dramatic loss in capability. This is particularly true for current generation reconnaissance satellites since these are very capable, relatively few in number, and very vulnerable owing to their need to be in low earth orbit. One option for mitigating this vulnerability is to deploy large numbers of less capable satellites. The commercial sector is now using such an approach to provide global cellular telephone capability. While the highest resolution imagery may still require large satellites, a network of small satellites could meet many needs and would provide graceful degradation in the event one is lost. In contrast to the loss of a single highly capable satellite, which could be crippling, the remaining

---

[118] Does United States Need Space Based Weapons ,William L SpacyII Major < USAF

satellites in a network would still provide significant capability.[119] Many countries like Britain, France and Israel lay stress in deploying Micro/mini satellites instead of large satellites to bring down the cost as well as the element of threat.

Responsive Space Lift Capability

Another step that can be taken to assure access to space is to develop a responsive space-lift capability. The ability to prepare and launch a satellite within days could quickly replenish combat losses. This approach would be most cost-effective for small, cheap satellites, but not for large satellites.[120] For the military roles India should design and develop smaller satellites as is being done in many countries. Large number of the small satellites can do many missions very well. These can be launched more cheaply and replaced more easily. They also have the advantage of graceful degradation under attack and present the enemy large number of low value targets each of which is difficult and expensive to destroy.

Offensive Counter Space

Jamming. An important objective of most conflicts has been disrupting the enemy's command and control system. While jamming the ground stations using large, highly directional antennas is difficult; the satellites themselves are much more vulnerable. The results of jamming these signals would vary from slow degradation of the orbit, to disrupting satellite communication networks but, these are difficult to verify. Another type of jamming is possible against reconnaissance satellites that take visual or infrared pictures is to track a satellite while it is overhead and train a laser on it. Even low-power lasers can temporarily blind optical sensors. The drawback of this approach is similar to that of jamming command signals: its effectiveness would be very difficult to verify. Since it would not be possible to tell if a satellite was actually blinded, there would be no way to

---

[119] Ibid

[120] Ibid

determine that the enemy remained unaware of troop dispositions or information was to be denied him.[121]

Wresting Control of Enemy Satellite. A potentially more effective approach may be to take command of an enemy satellite. It may be possible to break the codes used to command the manoeuvres of a satellite and send it spurious instructions. If transmitters are so placed that they could overpower legitimate commands, or send commands when the legitimate transmitters are out of range, then a satellite could be prevented from performing its mission. In contrast to jamming a satellite, the reactions of the satellite would make it possible to verify that the attack had been successful.

Direct Ascent Weapons. The most certain method of denying an enemy the use of his space-based assets is to physically destroy the satellites themselves. Ground-based directed energy weapons, KE ASATs, and co-orbital ASATs are the three major types of weapons. United States has pursued several anti satellite programmes in the past; it has actually intercepted an orbiting satellite with ASAT air-launched system which comprised of an F-15 fighter aircraft and a two-stage missile. The system was successfully tested against a satellite on September 13, 1985. The programme was cancelled in 1988 when congress voted to continue a two-year-old ban on further tests against objects in space.[122]

The US Army's KE ASAT programme is a ground-based two-stage missile that delivers a "kill vehicle" into the path of the target satellite using a derivative of the Minuteman booster, In Feb 2008, USA successfully intercepted a disused satellite to prove this capability in the field. China too earlier had carried out similar anti satellite test in Jan 2007 which added huge cloud of debris in the space.

---

[121] Does United States Need Space Based Weapons ,William L SpacyII Major < USAF
[122] Ibid

An alternative approach to an anti-satellite weapon is to match orbits with the target satellite and destroy it with an explosion in close proximity. The Soviet Union built and tested such a system during the 1970s and '80s. The system used either radar or infrared/optical sensors to home in on the target and exploded into a swarm of pellets when it came within range. Russia however, has forsaken the use of anti satellite weapons and along with China has sought an international agreement /treaty to ban weaponisation of the space.

Choice between Space Based and Ground Based Weapons. Ground based weapons have inherent advantages over their space-based counterparts moreover, deployment of space based weapons is forbidden by the Charter of the United Nation 1945 and various space treaties (the Partial Test-Ban Treaty 1963, the ABM Treaty 1972, the Environmental Modification Convention 1977 and Moon agreement 1979), including by the Outer Space Treaty 1967. Ground-based weapons have the ability to attack satellites in virtually any orbital inclination. If the launch site is located on the equator, then an ASAT will be able to engage satellites in any orbital plane. Ground based weapons have the additional advantage of being accessible for maintenance and modifications should they be necessary. It is also probable that ground-based weapons would cost less than their space based counterparts for obvious reasons. What would make a space-based system more expensive is the additional complexity needed for a weapon to survive months or years in orbit and then perform flawlessly. Ground-based weapons, on the other hand, could be stored in climate controlled buildings or silos until they were needed.[123]

Preventing Launch of Adversaries' Space-Based Assets. Virtually all space launches are made from fixed locations that are well known. The facilities needed to prepare the satellites and boosters for launch are highly vulnerable. A conventional attack on such a target will not only destroy the particular satellite, it will most likely cause severe damage to the launch

---

[123] Does United States Need Space Based Weapons ,William L SpacyII Major < USAF

complex and prevent or delay subsequent launches as well. To mitigate this vulnerability a potential foe could choose to create mobile launch systems. The sea-launch system being developed by an international consortium will be such a system and will provide the capability to launch from virtually anywhere on the world's oceans. However, it will also be quite vulnerable since it employs very large slow-moving vessels. A number of platforms, from attack submarines to land-based bombers could destroy these vessels with little difficulty. Needless to say that our own single satellite launch site despite having two independent launch pads and with third being constructed at the same location remains vulnerable, hence, alternative launch facilities need to developed and well dispersed and provided physical and aerial attack protection.

Attacking Anti Satellite Assets

Offensive Counter Space strategy is similar in concept to an Offensive Counter Air strategy. However, attacking ASAT weapons on the ground is less complicated than combat aircraft which are normally dispersed to many airfields in time of war. ASAT weapons will be deployed to relatively few easily identifiable launch sites. Once hostilities begin, these sites will be open to attacks by conventional assets such as stealth aircraft and cruise missiles.

Recommendations

No other space faring nation except India has shown antipathy to the military use of the space. As we move towards 21st century it is inevitable that space will become another medium of warfare. China near home is rapidly emerging as a space power and its capabilities may also be placed at the disposal of other countries for strategic and commercial reasons. Exploitation of the space would be great advantage against a non space capable nation such as enjoyed by Allied Forces in Iraq war and Afghanistan campaign but, advantage would be nullified against a space capable belligerent. In such scenario, the Favourable Space Control (Similar to Favourable Air Situation in context to Air Space) is feasible only if we retain our capability use the Space Assets while denying the use to the adversary This would also mean that we develop the capability to destroy / degrade the performance of enemy's satellites through various Anti Sat measures(direct hit method/ Directed Energy Weapons and active ECM etc) as well as attacking enemy space launch facilities on the ground which naturally will have to be planned. It is in this context India needs to address the role of space in our defence and national security. It is crucial that we incorporate existing space facilities, resources and operations in our security policies and strategy. While some dual-use facilities could be harnessed straight away, other capabilities would demand dedicated efforts.

Considering the complexity and multifarious space functions, the current space cell should be expanded progressively to a tri service space command in time bound manner with linkage with DoS/ISRO at an appropriate level. The next priority should be training of the requisite manpower at to handle space related equipment at various levels (command and field level). The manpower requirement would be huge considering the mammoth task. In US for example 27000 trained personnel (inclusive of trained civilian staff) perform the operations. For our kind of regional perspective, the scale of operation in our context is relatively small but still training of the adequate staff remains an important issue. Besides a separate budgetary support to meet the satellite requirements of the three services would also be needed.

Use task/mission specific cheap micro/nano satellites as their loss due to any reason (Enemy or un-serviceability) does not lead to total information blank out. Along with this ISRO should develop responsive space launch capability to replace satellite losses at short notice. We also need to develop multiple well dispersed launch facilities with adequate physical and electronic protection.

To fulfill offensive counter space requirements, we should develop ground based weapons like Directed Energy Weapons like lasers and electronic jammers to deter our adversaries. DRDO, having gained experience in the Ballistic Missile Defence (BMD) capability has claimed capability to deploy Anti sat weapon at short notice if operationally needed. Mere statement of the intention will not provide immunity to our space assets from destruction. The capability would act as deterrent only if this capability is demonstrated. Chinese Anti Sat test was criticized for creating enormous space debris which may take a millennium to burn out. As a responsible nation, we may carry out exo-atmospheric Anti Sat test at a lower height say in the height band of 100 to 150 miles so that while we display our capability but the ensuing debris burns out in about a week to ten days and does not become a cause of concern for the world community.

For secure network centric operations, satellites are planned for launch by individual service. The modality for collation, analysis and sharing of information also needs to be planned and practiced.

The launch of Indian INRSS by 2014 by ISRO should give us independent navigation and precision targeting facility. Our fighting elements in various mediums (air, sea and ground) should have requisite capability to operationally use GPS information. If this capacity is nonexistent/partially available then same should be built up progressively. We would also be able to use Russian GLONASS as we have invested in the equipment but, NAVSTAR GPS may not be available during hostilities more over the quality of information may not be of high order unless allowed by the Americans.

BIBLIOGRAPHY

1 Bhupendra Jasani, in a paper titled "Towards a future European space surveillance system " in Collective security in space published by Space policy institute , George Washington university.

2. Radha Krishan Rao in an article in "Chinese threat to Indian space assets "available at http://www.domain-b.com/aero/20090129_ indian_space.html.

3. Richard Kaufman Henry Hertzfeld in "Space security and Economy " Published by Economists for peace and security in September 2008.4. Raja Menon , in an article titled " Strategic Space " in Space security and global cooperation 2009 Published by Academic Foundation in association with Institute for Defence Studies and Analyses New Delhi.

4. Raja Menon , in an article titled " Strategic Space " in Space security and global cooperation 2009 Published by Academic Foundation in association with Institute for Defence Studies and Analyses New Delhi.

5. Boundary in http:/en.wikipaedia.org/wiki/oter_space.

6. Paris Convention 1914. Convention relating to the Regulation of Aerial Navigation signed at Paris on Oct 13, 1919 available at http://www.aviation.go.th/airtrans/airlaw/1914.html.

7. Convention on International Civil Aviation signed at Chicago on & Dec 1944. (Chicago convention) Available at http://www.mcgill.ca/files/iasl/chicago1944a.pdf

8. Treaty on principle governing the activities of the states in the exploration and use of outer space including the moon and other celestial bodies, available at http://www.state.gov/www/global/arms/treaties/space1.html

9. Dr. Raymond J. Barrett (Ph.D., Trinity College, Ireland) is Department of State Advisor, John F. Kennedy Centre for Military Assistance in Outer space and Air space: Difficulties in Definition .available at http://www.airpower.au.af.mil/airchronicles/aureview/1973/may-jun/barrett.html

10. United Nations Convention on The Law Of The Sea- available at http://www.un.org/Depts/los/convention_agreements/texts/unclos/closindx.htm United Nations

11. The Antarctic Treaty available at http://www.nsf.gov/od/opp/antarct/anttrty.jsp

12. United Nations Committee on the Peaceful Uses of Outer Space in http://www.oosa.unvienna.org/oosa/COPUOS/copuos.html

13. United Nations Treaties and Principles on Outer Space United Nations Treaties and Principles on Outer Space. Available at http;//www/oosa.Unvienna.org/oosa/spacelaw/treaties/html

14. K R Sridhara Murthy,V Gopalakrishanan and Parthasartha Sarthi Datta of Antrix Corporation Ltd "An article on." Legal Environment for Space activities"

15. Rajat Pandit. An article on "Dedicated Satellite for the Navy by the year End in Times of India of 20 May 2010.

16. The Freedom of Space doctrine, available at www. American foreign relations .com/o-w/outer space

17. Presidential Directive NSC-37 on US "National Space Policy "May 11, 1978 available at http://www.au.af.mil/au/awc/awcgate/nsc-37.htm

18. US National Space Policy, 31 Aug 2006 available at http://www.whitehouse.gov/sites/default/files/microsites/ostp/national-space-policy-2006.pdf

19. Bryan Johnson "The Military Use of Space "available at http://www.suite101.com/article.cfm/political_economy/19993#ixzz0oYl3SjyS

20. US Space Command Vision for 2020 available at www.fas.org/spp/military/docops/usspac/visbook.pdf

21. Background Paper: Peaceful And Military Uses Of Outer Space: Law And Policy Prepared By Institute Of Air And Space Law, Faculty Of Law, Mcgill University, Montreal, Canada February2005available at www.earl.net/.../space.../Background Paper%20McGill%20Outer%20Space%20Uses.pdf

22. U.S. Air Force Space Command, "Strategic Master Plan FY04 and Beyond," available in *www.fas.org/spp/guide/usa/StrategicMaster Plan04.pdf*

23. Soviet Military Space Doctrine available at www.fas.org/irp/dia/sovmilspace (Report produced by US government)

24. Theresa Hitchens "Developments in Military Space: Movement toward space weapons?"

25. Russian federal Space Agency at *en.wikipedia.org/wiki/Russian Federal_Space_Agency* –

26. The Military Doctrine of the Russian Federation approved by the Presidential edict on 5 Feb 2010 available at http://www.carnegieendowment.org/files/2010russia_military_doctrine.pdf

27. Squadron Leader KK Nair ," Space ,The frontier of modern defence 2006"

28. Space Debate available at   http://www.spacedebate.org/evidence/ 2211/

29. White Paper on China's National defence,www.china.org.cn/ government/central_government/2009-1/20/content_17155577_4.htm

30. China's Attitude toward Outer Space Weapons. Available at http:// www.nti.org/db/china/spacepos.htm

31. China , Space Weapons and US Security http://www.cfr.org/ publications/16707/

32. Mary c. Fitzgerald. " China's Military Strategy in Space " http:// www.hudson.org/files/publications/07_03_29_30_fitzgerald_ statement.pdf

33. Lisa Dome ," Chinese Space Policy Collaboration or Competition" , an Article published in Centre for Strategic and International Studies. http://csis.org/blog/chinese-space-policy-collaboration-or-competition.

34. Ashley J Telis, "China's Military Space strategy" available at www.carnegiceendows.org/files/tellis-china-space/.pdf

35. European Space Agency at http://en.wikipedia.org/wiki/ European_Space_Agency

36. A Defence Policy for Defence of Europe available at http:// www.eurodefense.de/images/aspacepolicyfordefenceofEurope.pdf

37. UK Space Agency available at http://www.bis.gov.uk/ukspaceagency/ what-we-do

38. Steven Berner ," Japans Space Programme : A Fork In the Road" A Rand Study

39. Japan's Basic Plan for Space Policy. Available at http:// www.kantei.go.jp/jp/singi/utyuu/basic_plan.pdf

40. Israel Space Agency available at http://www.globalsecurity.org/space/world/israel/agency.htm

41. Indian Space Research Organisation, http://www.isro.org/

42. Success of Cryogenic Rocket Launch will Make India a Leader in Rocketry Economic Times .Available at –articles.economictimes. indiatimes.com/keyword/cryogenic

43. Goals and Objectives available at http://en.wikipedia.org/wiki/Indian_ Space_Research_Organisation.

44. Report of the working group established by the steering committee of the planning commission 11th plan.

45. Isabelle Sourbes-verger, "Strategic space a variable geometry concept" in Space Security and Global Cooperation 2009

46. David Webb, Leeds Metropolitan University, On Definition of Space Weapon (When Space Weapon is not a Weapon?) http://praxis.leedsmet.ac.uk/praxis/documents/space_weapons.pdf

47. Joan Johnson Freese," Space as strategic Asset" Page 6 .2007, Columbia University Press

48. Jon Howland, JINSA Editorial Assistant, in *JINSA Online*, December 04, 2003. "Foes See U.S. Satellite Dependence as Vulnerable Asymmetric Target" in http://www.globalsecurity.org/org/news/2003/031204-jinsa.htm .

49. Debate on "Definition of the Space Weapon" in http://www.spacedebate.org/definition/Space%20Weapon/

50. Barry D Watts , Centre for Strategic and Budgetary Assessment , Feb 2001 "The military use of space , a diagnostic assessment" in http://www.csbaonline.org/wp-content/uploads/2011/02/2001.02.01-Military-Use-of-Space.pdf

51. Thomas. P. Ehrhard, Senior Fellow in Centre for Strategic and Budgetary Assessment, study "An Air Force Strategy for Long Haul" at http://www.csbaonline.org/4Publications/PubLibrary/R.20010201.The_Military_Use_o/R.20010201.The_Military_Use_o.pdf

52. "India Building a Military Satellite Reconnaissance System" in Defence Industry Daily, Aug-2005 at http://www.defenseindustrydaily.com/india-building-a-military-satellite-reconnaissance-system-0996/

53. ISRO To Launch Radar-Imaging Satellite on April20, 2009 a report at http://www.india-defence.com/reports-4320

54. Amitav Mallik, "Militarisation of Space : Security implications" Claws Journal distributed by Knowledge World , New Delhi Winter 2008 at www.claws.in/claws_journal_winter_2008.pdf

55. "Now a Space Cell to Keep an Eye on China Plans," A News report in Times of India of 11 Jun 2008.

56. India has anti-satellite capability: DRDO. A report in Thai Indian News at http://www.thaindian.com/newsportal/uncategorized/india-has-anti-satellite-capability-drdo_100318100.html

57. India Launches $200m TECSAR Spy Satellite. A Report in Defence Industry Daily at http://www.defenseindustrydaily.com/Indias-200M-TECSAR-Satellite-Purchase-Launched-05386/

58. Indian Regional satellite System, available at en.wikipedia.org/wiki/Indian_Regional_Navigational_Satellite_System

59. India and US 10 years Defence Pact. A Report in Defence Industry Daily 30 Jun 2005, available at http://www.defenseindustrydaily.com/us-india-sign-10year-defense-pact-0783/

60. Shri AK Dave , The Real Story of China's War on India ,1962(2006) in USI Paper.

61. US quits ABM Treaty, CNN News, Dec14, 2001 available at http://archives.cnn.com/2001/ALLPOLITICS/12/13/rec.bush.abm/

62. China's Defence Budget to Grow 7.5% in 2010 , A news report in China Daily, 4 Mar 10 available at http://www.chinadaily.com.cn/china/2010-03/04/content_9537753.htm

63. India vs China on Military Strength- Conventional and Nuclear available at http://www.abytheliberal.com/internationalism/india-vs-china-military-conventional-nuclear

64. Mr . Richard A Bitzinger , A Asia Pacific Security Study titled 'Civil Military Integration and Chinese Military Modernization," Volume3 Number 9, Dec 2004 available at   Http://www.apcss.org/Publications/APSSS/Civil- MilitaryIntegration.pdf

65. Ashley J. Tellis , Senior Associate, Carnegie Endowment for International Peace," Punching the U.S. Military's Soft Ribs: China's Anti-satellite Weapon Test in Strategic Perspective," The Policy brief no57,2007 is available at http://www.carnegieendowment.org/files/pb_51_tellis_final.pdf

66. Bharath Gopalaswamy and Harsh V. Pant , " Does India really need ASAT capability now?" The paper is available at http://www.defence.pk/forums/military-forum/44983-india-targets-chinas-satellites-3.html

67. Standing Committee On Defence (2003) ,(Thirteenth Lok Sabha) Nineteenth Report Ministry Of Defence, Http://164.100.24.208/Ls/Committeer/Defence/19.Pdf

68. An article in Times of India 06 Feb. 2007 titled" India away from setting up of Aero Space Command" available at   http://timesofindia.indiatimes.com/NEWS/India/India_years_away_from_setting_up_aerospace_command/articleshow/1565014.cms,

69. Defence Space Vision (DSV) 2020, available at http://mod.nic.in/aboutus/welcome.html

70. US Space Command, http://www.fas.org/spp/military/program/nssrm/initiatives/usspace.htm

71. THE STRATEGIC USE OF OUTER SPACE, The French White Paper on defence and national security, http://www.globalsecurity.org/military/library/report/2008/livre-blanc_france_2008-14.htm

72. White paper on China 's space activities 2006, http://www.spacedaily.com/reports/China_Issues_White_Paper_On_Space_Activities_999.html

73. People Liberation Army, http://en.wikipedia.org/wiki/People's_Liberation_Army

74. Military Uses of Space in POSTNOTE December 2006 ,Number 273. available at http://www.parliament.uk/documents/post/postpn273.pdf

75. British National Space Centre, available at http://en.wikipedia.org/wiki/British_National_Space_Centre

76. Lt Col Karl Ginter ,United States Army. USAWC Strategic Research Project 2006 titled " Space Technology and Network centric Warfare: A Strategic Paradox " available at http://www.google.co.in/search?hl=en&source=hp&biw=1259 &bih=620&q=Space+Technology+and+Network+centric+Warfare&oq =Space+Technology+and+Network+centric+Warfare&aq=f&aqi=g-v1&aql=&gs_sm=e&gs_upl= 18431l26561l0l1 00l23l0l1 1l11l6l2971331l510.16.3

77. Major William L Spacy, A research Project titled ," Does United States Need Space Based Weapons?", Published by Air University press available at http://aupress.au.af.mil/digital.asp?p=11

INDEX

www.ingramcontent.com/pod-product-compliance
Lightning Source LLC
Chambersburg PA
CBHW060421100426
42812CB00030B/3260/J

9 789380 177762